Studies in Big Data

Volume 107

The series "Studies in Big Data" (SBD) publishes new developments and advances in the various areas of Big Data- quickly and with a high quality. The intent is to cover the theory, research, development, and applications of Big Data, as embedded in the fields of engineering, computer science, physics, economics and life sciences. The books of the series refer to the analysis and understanding of large, complex, and/or distributed data sets generated from recent digital sources coming from sensors or other physical instruments as well as simulations, crowd sourcing, social networks or other internet transactions, such as emails or video click streams and other. The series contains monographs, lecture notes and edited volumes in Big Data spanning the areas of computational intelligence including neural networks, evolutionary computation, soft computing, fuzzy systems, as well as artificial intelligence, data mining, modern statistics and Operations research, as well as self-organizing systems. Of particular value to both the contributors and the readership are the short publication timeframe and the world-wide distribution, which enable both wide and rapid dissemination of research output.

The books of this series are reviewed in a single blind peer review process.

Indexed by SCOPUS, EI Compendex, SCIMAGO and zbMATH.

All books published in the series are submitted for consideration in Web of Science.

More information about this series at https://link.springer.com/bookseries/11970

Soumi Majumder · Nilanjan Dey

AI-empowered Knowledge Management

Springer

Soumi Majumder
Department of Business Administration
Vidyasagar University
Midnapore, West Bengal, India

Nilanjan Dey
Department of Computer Science
and Engineering
JIS University
Kolkata, West Bengal, India

ISSN 2197-6503 ISSN 2197-6511 (electronic)
Studies in Big Data
ISBN 978-981-19-0315-1 ISBN 978-981-19-0316-8 (eBook)
https://doi.org/10.1007/978-981-19-0316-8

This Springer imprint is published by the registered company Springer Nature Singapore Pte Ltd.
The registered company address is: 152 Beach Road, #21-01/04 Gateway East, Singapore 189721, Singapore

Preface

Our book is focused on AI-based Knowledge Management procedures following the modern trends of business practices. The purpose of the Knowledge Management process is to share perspectives, ideas, experiences, and information, to ensure that these are available in the right place at the right time to enable informed decisions, and to improve efficiency by reducing the need to rediscover knowledge. The objectives of knowledge management are to improve the quality of management decision making by ensuring that reliable and secure knowledge, information, and data is available through the service lifecycle, to enable the service provider to be more efficient and improve quality of service, to increase satisfaction and to reduce the cost of service by decreasing the need to rediscover knowledge. Knowledge Management (KM) is also a multidisciplinary field. AI allows machines to acquire processes and use knowledge to perform tasks and to unlock knowledge that can be delivered to humans to improve the decision-making process. AI has become the latest "buzzword" in the industry today. However, AI has been around for decades. The intent of AI is to enable computers to perform tasks that normally require human intelligence; AI is eventually evolving to take over many jobs once performed by humans. The connection of KM and AI has led the way for cognitive computing. Cognitive computing uses computerized models to simulate human thought processes. Cognitive computing involves self/deep learning artificial neural network software that uses text/data mining, pattern recognition, and natural language processing to mimic the way human brain works. Cognitive computing is leading the way for future applications involving AI and KM.

In recent years, the ability to mine larger amounts of data, information, and knowledge to gain competitive advantage and the importance of data and text analytics to this effort is gaining momentum. As the proliferation of structured and unstructured data continues to grow, we will continue to have a need to uncover the knowledge contained within these big data resources. Cognitive computing will be the key to extracting knowledge from big data. Strategy, process-centric approaches, and inter-organizational aspects of decision support to research on new technology and academic endeavors in this space will continue to provide insights on how we process AI with knowledge management to enhance decision making. Cognitive computing is the next evolution of the connection between AI and KM. AI can replace humans and explain what is important to consider in making the transformation to the digital organization of innovation. We conclude our study by exploring directions for future research.

In our book, Chap. 1 provides introduction to Knowledge Management. It says about various important parameters of knowledge management like types of knowledge, significance of knowledge management, knowledge management process in business, relation between knowledge management and information technology. Chapter 2 discusses the various tools of knowledge management that are used by business organizations in modern times. This chapter includes document management system, learning management system, customer relationship management system, decision support system, and social communication system. Chapter 3 elaborates knowledge management practice with the help of AI in different sectors. The sectors are like healthcare sector, construction sector, and education sector, small and medium-sized enterprises (SMEs), and e-business or e-commerce sector.

Chapter 4 is based on Knowledge Management System. It describes different types of knowledge management systems and their benefits. This chapter also gives an idea on how to build an effective knowledge management system in the organization citing few important examples of KMS.

Chapter 5 reports AI-empowered Knowledge Management. This includes importance and impact of Artificial Intelligence on Knowledge Management, different forms of benefits of AI for Knowledge Management, and the future of Knowledge Management.

Chapter 6 discusses the role of explainable AI (XAI) in Knowledge Management.

Bibliometric analysis of AI in Knowledge Management is reported in Chap. 7.

Finally, Chap. 8 concludes the book and summarizes the book content.

Kolkata, India

Soumi Majumder
Nilanjan Dey

Contents

About the Authors

Soumi Majumder completed her PGDM, specialization in Human Resource Management from All India Management Association (Institute of Business Management, Jadavpur University) in 2012. She did Diploma in Labour laws with Administrative laws (DLLAL) and Masters in Human Resource Management (MHRM) from Annamalai University, under Tamil Nadu Government in 2013 and 2019 respectively. Currently, she is a Ph.D. student of the Department of Business Administration at Vidyasagar University, Midnapore, West Bengal, India. She is associated with the School of Business, Sister Nivedita University, WB, India, Department of Business Administration, Netaji Subhash Engineering College, Kolkata. She is an associate researcher of Universidad International de La Rioja, Logroño, La Rioja, Spain. Previously, she was associated with Techno India College of Technology, NSHM College of Management and Technology, Dinabandhu Andrews Institute of Technology and Management, J. D. Birla Institute of Science and Commerce, Future Institute of Engineering and Management, West Bengal State Labor Institute, Siliguri, Department of Management Science, Sister Nivedita University, Kolkata, India, etc. She is having 6 years of experience in academia and 2 years of industrial experience. She has 20 research papers in national, international conferences and journals in the area of quality work-life, decent work-life, work-life balance, stress management, employee engagement, job satisfaction, leadership, training, and learning, etc. She has presented a research paper at DSMLAI 2021 organized by India–Namibia Centre of Excellence in Information Technology, Namibia University of Science and Technology, Namibia. Furthermore, she is a member of the All India Management Association, ACM, IS, NIPM etc.

Nilanjan Dey is an Associate Professor in the Department of Computer Science and Engineering, JIS University, Kolkata, India. He is a visiting fellow of the University of Reading, UK. He also holds a position of Adjunct Professor at Ton Duc Thang University, Ho Chi Minh City, Vietnam. Previously, he held an honorary position of Visiting Scientist at Global Biomedical Technologies Inc., CA, USA (2012–2015). He was awarded his Ph.D. from Jadavpur University in 2015. He is the Editor-in-Chief of the *International Journal of Ambient Computing and Intelligence* (IGI

Global). He is the Series Co-editor of *Springer Tracts in Nature-Inspired Computing* (SpringerNature), Data-Intensive Research (SpringerNature), *Advances in Ubiquitous Sensing Applications for Healthcare* (Elsevier). He is having 110 books and over 300 publications in the area of medical imaging, machine learning, computer aided diagnosis, data mining, etc. He is the Indian Ambassador of the International Federation for Information Processing—Young ICT Group and Senior member of IEEE.

Chapter 1
Introduction to Knowledge Management

1.1 Knowledge Management

The management consulting community has given birth to the concept and terminology of knowledge management and has introduced it slowly in the business. In the past, when the internet came to the fore, organizations realized that intranet is one type of in-house subset of the internet. It was an excellent tool by which information could be accessed and shared easily within geographically dispersed organization units. They realized that to market a new product they should build tools and techniques such as dashboards, best practice databases, expertise locators, and so on. The term' knowledge management was first introduced in 1987 in the current context of McKinsey [1]. James. O. McKinsey was the founder of an American worldwide management consulting firm. It is the world largest strategy consulting firm by revenue. McKinsey focused on the complex tasks of the organization. In order to improve company performance, he emphasized the issues that affect much toward the company's performance. In this way, he developed and implemented strategies for the business to grow and develop over time. The other vital functional areas were implemented, i.e., projects in sales marketing, logistics and manufacturing, corporate Human Resources (HR), and finance and information technology. It was introduced due to an internal study on their information handling and its optimum utilization. After that, Knowledge Management (KM) became public at a conference in Boston in 1993 [2]. The famous definition of knowledge management was given by Tom Davenport in 1994 [3]. According to him, knowledge management is the technique to acquire, disseminate and utilize knowledge effectively. Knowledge management is the process by which organizations can analyze and use the impact of the collective knowledge of a group. In the world of business, it has been said that knowledge management is the method for the maintenance of a knowledge base or a portal. The specific knowledge of the companies is housed in this portal. Knowledge management helps businesses and employees to break and access information appropriately. This system provides a platform for the working people to impart knowledge that they have acquired over time; it also prevents the loss of information of a business

S. Majumder and N. Dey, *AI-empowered Knowledge Management*,
Studies in Big Data 107, https://doi.org/10.1007/978-981-19-0316-8_1

when individuals leave the organization. The features of knowledge management are (i). it captures and organizes knowledge to address particular kinds of business tasks as well as special projects, (ii) it helps to share knowledge with others who can be benefited, (iii) to improve processes and technology for providing easy access to knowledge, and finally, (iv) knowledge management promotes the platform for new generation knowledge for continuous learning [4].

Every organization must use technology to accomplish its challenges and goals. One of the central challenging tasks for the managers is to manage the knowledge of the organization. The range of knowledge workers lies within marketing, finance, and HR professionals to project managers, business analysts, and software engineers. They consume and generate knowledge daily because knowledge is the prime component of their responsibilities. For a dynamic and successful organization, knowledge always plays a vital role in management practices. The process of knowledge management includes all kinds of activities about the creation, distribution, and maintenance of knowledge to achieve the organizational objectives. In this scenario, HR management is the key to ensuring knowledge management in the organization. Organizations are conducting design on Human Resource Management System (HRMS) [5] to overcome the traditional hurdles in knowledge management. The most significant parameters under knowledge management are as follows [6]:

1. Promotion of knowledge contribution
2. Identification of sources of knowledge
3. Replenishment of appropriate knowledge
4. Protection and preservation of knowledge.

It is the modern practice in organizations where saving knowledge in an easily accessible form can improve organizational efficiency. The main goal of knowledge management is to keep the correct information in front of someone at an appropriate time and a just place.

1.2 Importance of Knowledge Management

A very famous line said by Paulo Coelho is, that, if there is no transformation of knowledge there is no wisdom. One of the greatest legacies of a company is knowledge. It helps to generate value for the business and allows the business to make assertive decisions. It has been seen in reality that many companies have a lacuna in organizing and storing information. Firms have faced a constant challenge in the form of collective learning and systematic usage of acquired knowledge. Therefore, knowledge management has a significant importance in business environment. It is established that organizational efficiency level depends on its decision-making ability. Knowledge management procedure boosts the technique of problem solving and decision making. A smarter workforce always gives quick decisions which are beneficial for the organization. An organization shall not only focus on high customer benefits to gain benchmarking in the industry but also should emphasize on reduction in attrition and generate innovation in the business processes to maintain the organization. Over time, the need for knowledge management in business is developing to make the business more competitive. Organizations are providing high emphasis on building and maintaining flexible and creative methodology. Information and innovation both can be accessed from any distance with the help of proper knowledge management. There are different reasons to establish an effective knowledge management process. The reasons are: (i) a merger and acquisition can encourage team members to share their expertise through knowledge management system, (ii) there is a need to capture knowledge from the key employees of the organization who are about to retire, (iii) knowledge management also helps in the induction programs for new employees and (iv) a recruitment drive can manage and store wisdom through utilization of appropriate knowledge management. Poor transformation of knowledge will create less efficiency and productivity in the business. By increasing availability and accessibility of the company's accurate knowledge, an effective knowledge management system can reduce the cost of inefficiencies. When the company has multiple branches and needs to hire remote employees, they must emphasize distribution and development of knowledge to accomplish the firm's objectives. Given this scenario, a different benefit of knowledge management has been discussed below.

1.2.1 Recreating Existing Knowledge Spending Less Time

Knowledge management reduces interruptions of co-workers in each other's workspace by e-mailing, chatting, faxing, etc., when accurate and feasible information is required. Team members and other employees spend less time answering repetitive questions. They can focus more on high-margin profit-yielding work. More efficiently managing of organizational knowledge will result in better usage of existing knowledge.

1.2.2 Get the Information Faster

Proper and accurate knowledge management helps businesses to get information faster. It becomes easy to jot down information quickly rather than asking or sending mail to different departments in the organization for necessary information. People who do not know the related and required information tend to ignore emails, resulting in unproductive communications. Futile circulations of emails are carried out as insignificant outcome in the form of appropriate information is obtained. An appropriate knowledge management practice reduces dependency on the functional units of the organization regarding information collection.

1.2.3 Make Fewer Mistakes

Improper transfer of correct information between employees results in error-prone deliverance of duties and showing inept attitude toward their responsibilities. This leads to an undesirable termination of knowledge transformation creating an adverse impact on individual productivity. Importance of knowledge sharing can be realized if learning from past mistakes is incorporated into the employees' outlook. In this way, knowledge management leads to a platform for reduced/limited mistakes.

1.2.4 Make the Decisions Informed

Through the knowledge management practice, employees can share their experiences, newly learned lessons, research, and searchable knowledge; other people of the organization are able to get access and can review this information. Multiple pieces of data and different viewpoints on particular topics are easy to gather, and based on that, decisions can be made for the firm's benefit.

1.2.5 Provide Better Service to Customers and Employees

Effective knowledge management provides support to team members of any particular functional project and helps to resolve problems. This provides solutions to the customer in case they face any problem related to the products or services of the business. Customer requests can be properly catered to within no time. With proper knowledge management, employees can experience a contented frame of mind and therefore become more productive at their jobs. On the other hand, trust level of customers increases toward the company resulting in more business dealings of the company's goods or services. The company can earn reputation, recognition, and revenue in the market. Short-term and long-term goals are achieved effortlessly.

1.2.6 Standardization of Processes

Knowledge management introduces standardization of processes in the firm. The standard and approved process which is established by top-level management should easily be accessible to everyone with the help of the unique practices in the workplace. It is easy to build records and information sharing processes are relatively easier. For example, in case of telephone games, we know that information gets distorted through word-of-mouth communication. A standard process of communication and information sharing is always needed. Accurate knowledge management makes the processes standard and authentic.

Therefore, it is crucial for the company's management that apart from focusing only on forming strategies about storing and sharing knowledge, they should also encourage their employees to do the same. To be successful, every company needs to create a culture of knowledge sharing and transformation in the work environment. The goal of any business organization should not only depend on the positive business outcomes; rather it must emphasize and encompass positive outcomes for the workforce also. In due course of time, when employees gather the positives of knowledge environment, automatically, the practice of knowledge management program will gain impetus in the business house.

1.3 Types of Knowledge

Knowledge management helps organizational people to learn and grow. The knowledge can have various forms and it has an impactful application in the practical domain of work as well. The concept of different kinds of knowledge has been introduced slowly to make the business more authentic and powerful. There are four types of knowledge such as implicit, explicit, tacit, and embedded. Knowledge has a significant impact on business functions. Knowledge management can use application of different knowledge to brief the way which can make the business more resourceful. Among others, explicit and tacit knowledge is more useful. Following is the discussion of different types of knowledge.

1.3.1 Explicit Knowledge

Explicit knowledge is codified and formalized in nature. Explicit knowledge is also mentioned as know-what, easy to identify, store and retrieve. Explicit knowledge can be easily handled by knowledge management systems (KMS), which help store, retrieve, and modify documents/ texts [7]. The most significant issue with explicit knowledge, comparable to information, is from a managerial standpoint. It entails ensuring that people have access to what they require, that essential knowledge is stored, and that knowledge is examined, updated, or destroyed as necessary. Explicit knowledge is basic and vital. It is a written form of knowledge with high accessibility. In this form of knowledge, data can be processed, organized, structured, interpreted, and implemented in the workstation. Explicit knowledge can be easily communicated, articulated, recoded, and most significantly stored in knowledge management. Knowledge management platform is an example of explicit knowledge which includes company white papers, datasheets, reports on research, etc.

1.3.2 Implicit Knowledge

Implicit knowledge is the practical application of explicit knowledge. Instances of implicit knowledge can be easily observed in an organization. Knowledge management encompasses implicit knowledge, which also is crucial for the workplace. For instance, when an executive of the company asks a team member about how to perform a task; various thoughtful processes can come out from their conversation. This discussion can generate potential outcomes, and lead to the best course of action to determine the performance. This is the application of implicit knowledge of a team member where it becomes prominent how to do a particular task efficiently and what can be the probable outcome. In addition, best practices, skills, and expertise that are carried forward from one task to another are also examples of implicit knowledge.

Implicit knowledge gained through explicit knowledge has the power of application to a specific type of situation in the workplace. For instance, if a book on flight mechanics and an airplane cockpit layout is one type of explicit knowledge, then implicit knowledge is the application of that information, that is, how to fly the airplane. The best method to learn something and gain knowledge is known as implicit knowledge. Synthesis of implicit knowledge helps in quick resolution of an entirely new problem. Implicit knowledge, not being scalable is, therefore, excluded traditionally from formal knowledge.

1.3.3 Tacit Knowledge/Embodied Knowledge

Tacit knowledge is known as embodied knowledge which refers to know-how. This type of knowledge is majorly experience-based knowledge and is hard to explain. Tacit knowledge is personal in nature and context-dependent. This knowledge is deeply related to action, involvement, and commitment and is difficult to communicate. The most valuable source of knowledge in the workplace is tacit knowledge. If there is a lack of tacit knowledge, it results in reduced capability of innovation and sustainability in competition. Tacit knowledge cannot be handled or managed by knowledge management systems. Tacit knowledge is the application and implementation of implicit knowledge in the workplace, under the discretion of the company. Due to attrition/recruitment in a company, application of embodied knowledge or tacit knowledge gained during the tenure changes. Tacit knowledge is unique for every company/business. For instance, when a salesperson gives a demonstration about a product to the customers or prospects, it showcases his expertise in imbibing specific buying signs and symbols while dealing with customers or prospects. Ultimately, it is not that significant as to how the complete knowledge of the company is defined. This knowledge plays a vital role in day-to-day business operations and helps in running the business efficiently. It is very important for a company to know and realize how to store and communicate different types of knowledge, effectively. This is required to build a knowledge-sharing strategy.

An iceberg metaphor can be used to describe tacit and explicit knowledge (Fig. 1.1). The visible tip of the iceberg is explicit knowledge [8]. The knowledge that can be written down and documented in a form that can be seen and read is the tip of the iceberg (5%, data, information, files: easy to be written down). But there are other things besides this virulent, codified form of knowledge. There is invisible knowledge. The rest of the iceberg is hidden underwater. That is tacit knowledge (95%, experience, thinking, competence, commitment: difficult to transfer). The knowledge in our heads is less specific and more challenging to document. These two forms of knowledge exist simultaneously, and all knowledge is based on these two forms.

Table 1.1 illustrates the difference between explicit knowledge and tacit knowledge/embodied knowledge.

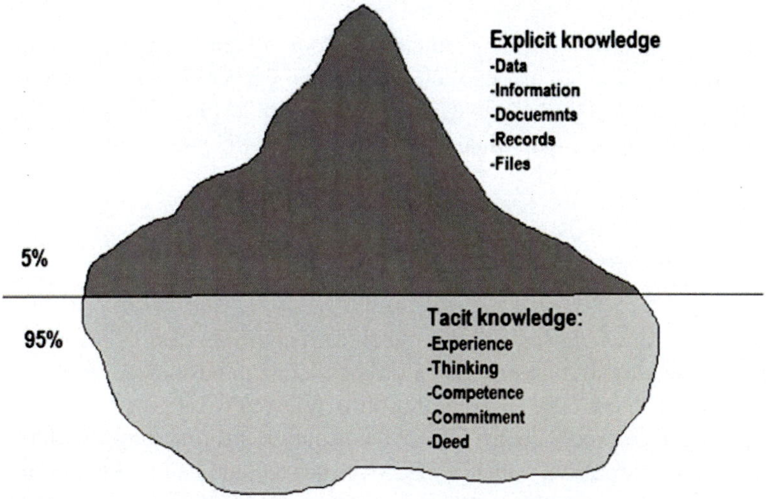

Fig. 1.1 The "iceberg" metaphor—tacit and explicit knowledge relationship

Table 1.1 Explicit knowledge versus tacit knowledge/embodied knowledge

Explicit knowledge	Tacit knowledge/embodied knowledge
Formal	Personal and context-specific
Tasks and procedures	Knowledge from memory
Codified	Experimental
Easy to document, transfer and reproduce	Difficult to document and communicate
Reports, processes, policy manuals (examples: online tutorials, policy and procedure manuals)	"Know-how" (examples: hands-on skills; special know-how; employee experiences)

1.3.4 Embedded Knowledge

Embedded knowledge is used in rules and regulations, products, processes, manuals, values and ethics, cultures, artifacts, codes of conduct, routine, and structures. Knowledge can be embedded either formally or informally. Formal embedded knowledge is like an initiative that has been taken by management to formalize a specific healthy routine. On the other hand, informally embedded knowledge is used when organizations utilize and apply the other two types of knowledge. It is essential to mention that sometimes embedded knowledge can exist in explicit sources though the knowledge is not explicit (i.e., a rule can be written in a manual). Culture and routines are two phenomena that are hard to change over time and difficult to understand. It is observed that the formalized routines are easier to implement in the workplace. Management can embed the fruits of lessons with the procedures, products, and routines.

A successful knowledge management initiative influences the conversion of tacit and embodied knowledge into explicit knowledge. Proper documentation of knowledge helps to enhance the efficiency and productivity level of the firm. It eliminates dependency on selected individuals in the workplace when they are not available.

1.4 Knowledge Management Process

The process of knowledge management includes capturing, storing, organizing, verifying, securing, distributing, and using of knowledge in business. The way of entire management of knowledge in organization is referred to as process of knowledge management. This process helps to provide accurate information, stimulate collaboration, create innovation, enhance effective internal and external communication. Managing knowledge in the organization also gives an enhancement in decision-making ability. In case of developing knowledge management process in a business environment, it requires an initial assessment of existing business process. Therefore, this becomes easy for the team members so that they can integrate the additional steps of knowledge management which will make a proper sense in the process. The paring in between right steps of knowledge management and appropriate business process can give the birth of high accessibility of knowledge in workplace; it gives ease to the working people to make important decisions. Apart from that knowledge management trains employees and increases their expertise. There are seven steps in knowledge management process. The steps are discussed below.

1.4.1 Discovery

Discovery of knowledge is the stepping stone of the knowledge management process, which signifies extraction of information from data. This information is essential for formulation of an organization's strategy and it also eases the process of communication, various business operations, and development of relationship. Data mining is helping to identify trends, patterns, and aids in correlation with customer relationships. Consequently, discovering more data is becoming easier.

1.4.2 Capture

The second step of knowledge management is capturing pre-processed internal and external knowledge. This knowledge is documented, communicated, and shared with others for benefits of business. Existing documents are required to be audited in this step. Content creation in knowledge area is encouraged through appropriate knowledge capturing. As a result, reduction of key gaps in knowledge management process can be attained.

1.4.3 Organization

Organization, not the business house, here it means a proper description, classification, categorization, and index of information, is the third step of knowledge management process. In this manner, organization of knowledge is retrieved, reused, navigated, and shared among employees, team members, and other critical users, efficiently. Proper organization of knowledge helps to sort and segment knowledge and therefore the necessary information is accessible by the users with high demand.

1.4.4 Assessment

The organization must ensure that knowledge should be verified and validated. Therefore, assessment has a principal role in knowledge management process. Significance of assessment is more because knowledge is associated with decision-making ability, spontaneous collaboration, innovation and creativity, and improvement of internal and external processes. Knowledge should be accurate, consistent, complete, and updated with the integration of other stages of knowledge management process. Regular reviews by external experts are extremely critical at this juncture.

1.4.5 Sharing

Knowledge sharing becomes very significant and is the key drive for people who actively seek knowledge in the organization. It helps organization to make knowledge available for everyone. Potential employees have the ability to implement and use knowledge in a better way, so that knowledge sharing helps in enhancing their understanding and therefore their expertise. It provides various advantages for the business. Knowledge sharing process can also fulfill the short and long-term objectives of the business. Team leaders must encourage their team members to share knowledge among themselves for learning. Team members should be rewarded with incentives to boost up the process.

1.4.6 Reuse and Application

The next level of knowledge management process is reuse and application of knowledge. When information is properly captured, organized, and shared among people of the organization, it generates techniques of reuse and application of this knowledge. Knowledge management increases the efficiency of the business. It provides significant improvements in the area of business operations by completing various tasks with strategies. The communication process between colleagues, customers, clients becomes more effective and transparent. Reuse of knowledge and its implementation has high demand in business in order to fulfill all actions. Interaction between employees and clients plays an important role in measuring effectiveness of reuse and application of knowledge process.

Relation between discovery, capture, sharing, and application [9] is clearly shown in Fig. 1.2. Combination encompasses facilitating the discovery of new explicit knowledge whereas socialization is the process that aids the discovery of new tacit knowledge. The process by which knowledge is converted from tacit to explicit

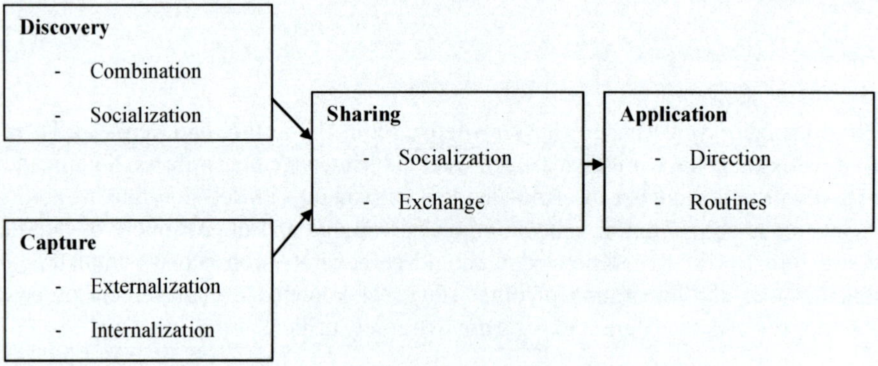

Fig. 1.2 Knowledge management process

form (contained within people, artifacts, or organizational entities) and vice versa, through the sub-processes of externalization and internalization, is termed as knowledge capture. The knowledge being captured is not restricted to being contained within the organizational boundaries. It can reside within external entities such as consultants, competitors, customers, suppliers, etc. When available knowledge is used to take decisions and perform tasks through direction and routines, then it is termed as knowledge application.

1.4.7 Creation

The final stage of knowledge management process is creation of knowledge. When the workforce of an organization has already gained the know-how through internal and external interactions, process of navigation, and high practice, then they are able to gather collective knowledge. Individual knowledge can be shared among employees, and then it is reused and applied. It is then finally expanded in the near future for knowledge seekers. Content enhancement for organizational guidelines and strategy making can help in establishing a culture of knowledge creation for all employees in the workplace. Significance of the process should be increased by praising and rewarding existing knowledge seekers. Communication process between team leaders and members is very effective in this situation and has an impact on knowledge creation.

1.5 Knowledge Management and Information Technology

In the field of knowledge management, information technology [10–15] provides a wide range of operations such as communication, coordination, browsing on storage and retrieval, presentation, numerical computation, location, filtering, symbolic processing, video conferencing, reasoning, e-mail, database management, software scheduling, hypertext, retrieval of information, internet and intranet, GUI, software for statistical analysis, web browser, presentation software, expert system, intelligent agents, artificial intelligence and many more. Socialization includes interaction between two or more persons who have a similar kind of interest or otherwise. IT supports functions that relate to communication, coordination, and group processes which is very useful in the process of socialization. There are many organizations in the market that publishes yellow pages which list experts and their expertise. Yellow pages use storage and retrieval functions.

On the contrary, the process of externalization converts tacit knowledge into explicit knowledge. This process can be accentuated by knowledge acquisition tools and techniques. The combination process creates a high quantity of explicit knowledge through transformation, analysis, and integration of less available explicit knowledge in the workplace. All the IT functions, except communication and coordination, are very much helpful in facilitating combination. In this context, it can be said that explicit knowledge is converted into tacit knowledge by internalization. Management of explicit knowledge can be done through significant investment in information technology [16–18]. In the organizational knowledge base, many things are captured like policies and procedures of the organization, problem-solving agendas, etc. The workforce is becoming more knowledgeable by the usage of browsing, filtering, presentation, location functions. In the organization, an expert system should be used as a tool for reusing knowledge. In this way, information technology is playing a vital role in the era of knowledge innovation, creation, and management.

The relationship between knowledge management and information technology is similar to tools and the mind. For example, mind can use the tools to transfer ideas and then explain. Similarly, knowledge management uses technology for knowledge transfer and exchange. Therefore, whenever any technological development takes place in the business, it influences various types of knowledge management processes. In recent years, increase in the effectiveness of knowledge management by development of information technology and its applications, has been observed. The contribution of information technology toward knowledge management is discussed below [19, 20].

1.5.1 The Role of Computers and Servers' Technology in Knowledge Management

In the modern economy, businesses use computers and therefore can store a large quantity of data and information. Usage of modern technology produces realistic and accurate knowledge in business. Computers and servers are essential in knowledge management as they have massive storage capacity and a large amount of information can easily be preserved. Whenever required, the information can be easily accessed. The speed of processors and their accuracy have a relation with modern servers and computers due to their ability to process large amounts of data quickly and correctly in record time.

1.5.2 The Role of Communications and Network Technologies in Knowledge Management

The development of communications and digital network technologies has spread out and made knowledge available for all. Internet is an appropriate example of communication and networking technology. Communication and networking technology plays a significant role in knowledge management.

Fig. 1.3 Knowledge management in the digital world through the Internet of things (IoT)

1.5.3 The Role of Internet Technology in Knowledge Management

Internet technologies have changed the traditional functions of a business. Internet-based technology provides information to the people of the organization and makes them knowledgeable. Furthermore, it transfers knowledge among others. The application of Internet of Things (IoT) [21, 22] is rapidly developing and spreading. Presently, different devices are connected to the internet [23] and aid in revival and exchange of information (Fig. 1.3.). This makes it easier for knowledge management to collect data directly from equipment and devices. In this way, knowledge becomes more accurate and effective. Therefore, it shows that the role of internet technology is also significant under the domain of knowledge management.

1.5.4 The Role of Image and Video Technologies in Knowledge Management

During past decades, there has been growth and development in the usage of cameras and scanners [24]. This acts as one of the essential sources of knowledge management. It helps to collect and analyze information. Identification of patterns, recognition of details, and comparison of images has been made easier than before through camera technologies and software packages. With the help of OCR technology, it is possible to recognize numbers and letters. Here scanner technology and its software packages are frequently used. The image and video technologies make it easier to receive and store information in knowledge management [25–28].

1.5.5 *The Role of Printing Technology in Knowledge Management*

Printing technology plays an important role in knowledge management. Nowadays, various printing techniques are being developed, which helps knowledge management to circulate and disseminate knowledge more effectively. The example, in this case, is like 3-D printing technology which helps knowledge management to show information on products and services very clearly and efficiently.

1.6 Conclusions

It is very much true and known to all business leaders that an organization's asset management is one of the prime conditions of a successful business. When it comes to the most valuable asset management, many organizations along with the combined knowledge of their employees are seen to struggle. The most valuable asset of an organization is the combined knowledge of its employees. By effectively managing these assets, the organization can achieve success. Sometimes, the company leaders overlook the need for knowledge management in business, which then provides a negative impact on business effectiveness. Business organization is such a place where numerous employees are working together and knowledge sharing among them has become one of the key criteria for knowledge enhancement. Knowledge can be gained from experienced colleagues who help in increasing skills and expertise. A proper way of knowledge management creates a structured and effective environment that captures and spreads useful knowledge at its best. There are so many advantages when knowledge management becomes an impactful practice in an organization's climate. It emphasizes costs–benefit analysis, improves decision-making ability, creates smooth workflows, makes scalability easier, reduces and prevents brain drain, generates a high involvement of knowledge sharing culture among human resources. Knowledge should be accurate, dynamic, and personal in nature. Knowledge management has a deep relation with artificial intelligence. Machine learning techniques allow machines for accruing, storing, processing, and using knowledge to perform tasks and duties. It unlocks knowledge that is distributed among the people for improving decision-making process. In the last few years, knowledge management has created an image of problem-solving in critical areas of the business functions. Therefore, it can be said that the philosophy of the organizations has become more collaborative and team-oriented. Many collaboration tools have been discovered by the science of management. In corporations, these innovative tools are the key drivers for knowledge sharing and information transformation. Knowledge management practice leads toward growth and development of the organization. It gives a positive picture of the company.

References

1. Maier, R., & Hadrich, T. (2011). Knowledge management systems. In *Encyclopedia of knowledge management* (2nd ed., pp. 779–790). IGI Global.
2. Demarest, M. (1997). Understanding knowledge management. *Long Range Planning, 30*(3), 374–384.
3. Wiig, K. M. (1997). Knowledge management: An introduction and perspective. *Journal of knowledge Management*.
4. Alavi, M., & Leidner, D. E. (2001). Knowledge management and knowledge management systems: Conceptual foundations and research issues. *MIS Quarterly*, 107–136.
5. Varun Grover, T. H. D. (2001). General perspectives on knowledge management: Fostering a research agenda. *Journal of Management Information Systems, 18*(1), 5–21.
6. McElroy, M. W. (2010). *The new knowledge management*. Routledge.
7. https://www.galaxyconsulting.net/7-blog/135-knowledge-types-and-knowledge-management. Last access date: 9 Oct 2021.
8. https://ryhma5blog.wordpress.com/2016/10/18/tacit-knowledge/. Last access date: 9 Oct 2021.
9. Becerra-Fernandez, I., & Rajiv, S. (2010). Knowledge management: *Systems and processes*. M. E. Sharpe. Last access date: 9 Oct 2021.
10. Sabherwal, R., & Sabherwal, S. (2005). Knowledge management using information technology: Determinants of short-term impact on firm value. *Decision Sciences, 36*(4), 531–567.
11. Mohamed, M., Stankosky, M., & Murray, A. (2006). Knowledge management and information technology: Can they work in perfect harmony? *Journal of Knowledge Management*.
12. Okumus, F. (2013). Facilitating knowledge management through information technology in hospitality organizations. *Journal of Hospitality and Tourism Technology*.
13. Holtshouse, D. K. (2013). *Information technology for knowledge management*. Springer Science and Business Media.
14. Tseng, S. M. (2008). The effects of information technology on knowledge management systems. *Expert Systems with Applications, 35*(1–2), 150–160.
15. Tanriverdi, H. (2005). Information technology relatedness, knowledge management capability, and performance of multibusiness firms. *MIS Quarterly*, 311–334.
16. Bahramimianrood, B., & Bathaei, M. (2021). The impact of information technology on knowledge management in the supply chain. *Journal of Social, Management and Tourism Letter, 2021*, 1–11.
17. Rusilowati, U., Supratikta, H., & Metarini, R. R. A. (2021). Innovation of government research and development institution based on knowledge management and information technology (case study on the Government Policy-Making Research and Development Institution). *Budapest International Research and Critics Institute (BIRCI-Journal): Humanities and Social Sciences, 4*(3), 4348–4362.
18. Al Mansoori, S., Salloum, S. A., & Shaalan, K. (2021). The impact of artificial intelligence and information technologies on the efficiency of knowledge management at modern organizations: A systematic review. *Recent Advances in Intelligent Systems and Smart Applications*, 163–182.
19. Lam, L., Nguyen, P., Le, N., & Tran, K. (2021). The relation among organizational culture, knowledge management, and innovation capability: Its implication for open innovation. *Journal of Open Innovation: Technology, Market, and Complexity, 7*(1), 66.
20. González-Díaz, R. R., Acevedo-Duque, Á. E., Flores-Ledesma, K. N., Cruz-Ayala, K., & Guanilo Gomez, S. L. (2021, March). Knowledge management strategies through educational

digital platforms during periods of social confinement. In *World Conference on Information Systems and Technologies* (pp. 297–303). Springer.
21. Shinde, G. R., Dhotre, P. S., Mahalle, P. N., & Dey, N. (2021). *Internet of things integrated augmented reality*. Springer.
22. Mahalle, P. N., Shinde, G. R., & Deshpande, A. V. (2021). *The convergence of Internet of things and cloud for smart computing*. CRC Press.
23. https://www.linkedin.com/pulse/increasing-importance-knowledge-management-digital-world-kgabo-badimo/. Last access date: 1 Oct 2021.
24. Yu, Z., Patnaik, S., Wang, J., & Dey, N. *Advancements in mechatronics and intelligent robotics*.
25. Babo, R., Dey, N., & Ashour, A. S. (Eds.). (2021). *Workgroups eAssessment: Planning, implementing and analysing frameworks*. Springer.
26. Dey, N., & Santhi, V. (Eds.). (2017). *Intelligent techniques in signal processing for multimedia security*. Springer International Publishing.
27. Dey, N., Ashour, A., & Acharjee, S. (Eds.). (2016). *Applied video processing in surveillance and monitoring systems*. IGI Global.
28. Jat, D. S., Bishnoi, L. C., & Nambahu, S. (2018). An intelligent wireless QoS technology for big data video delivery in WLAN. *International Journal of Ambient Computing and Intelligence (IJACI), 9*(4), 1–14.

Chapter 2
Tools for Knowledge Management

2.1 Introduction

It is well established that knowledge management is not a single domain. Instead, knowledge management is an integration of various endeavors and a unique field of study. Presently, knowledge management practitioners benefit from using Knowledge Management (KM) tools and techniques in a business organization. This has provided an overview of a list of concepts and critical terms. Knowledge management tools are unique technologies that can enable and enhance knowledge creation, codification, and transfer within business firms. This KM tool reduces the burden of work and allows application of resources and effective usage toward practical tasks [1]. The tools guide practitioners to describe various frameworks and explore several potential application areas. This also facilitates them how to use KM tools efficiently. It is essential to state that all knowledge tools are not computer-based; paper and pen also can be used to generate, codify and transfer knowledge in a business environment. In due course of time, the knowledge tools have been accentuated with technologies. Tools are created for quick evolution of business processes; enhance dynamic capabilities and effectiveness of the organization. Knowledge tools are the most expensive and worthy in true sense. The pressure of knowledge management tools and techniques are related to communication and collaboration tools for sharing knowledge. Knowledge management tools transfer the process into practice. The growth of information technology has created so many ERP (enterprise resource planning) type systems that can easily manage a well-defined process into business [2]. Knowledge management practitioners or experts utilize a wide range of information technology tools for sharing, creating, and codifying the knowledge. The high growth of information technology leads to more communication and collaboration tools. The tools created for knowledge management have been used for comprehension, assimilation, and learning of the information by individuals working in the company. Then they will be able to transform the data and information into knowledge. Knowledge is essential for individuals and groups, and people support

S. Majumder and N. Dey, *AI-empowered Knowledge Management*,
Studies in Big Data 107, https://doi.org/10.1007/978-981-19-0316-8_2

knowledge management tools to get easy availability and accessibility of information systems. In this way, the central and seen part of knowledge, known as explicit knowledge in literature and opposed to tacit knowledge dimension, becomes so much significant under knowledge management. Therefore, there is high requirement for knowledge management tools in business organizations that help transfer knowledge into various forms and types of documents and media [3].

2.2 Document Management System

The document management system is an essential and valuable tool of knowledge management. It has a significant impact on the functions of business organizations. When it comes to the domain of organizational activities, all the functional departments of the business house must focus on document management [4]. A document is such a thing that can be recorded in written, photographed, or secured in some other form. The document management system is an automated software solution that helps to organize, secure, capture, digitize, tag, approve, and complete the tasks of files and documentation. Mostly, document management systems store data in the cloud. A document management system works more than cloud storage. We can consider the example of the e-File cabinet under an advanced document management system. This technology handles a large number of papers that flow into the business. On the other hand, it can be said that document management is a special kind of system that captures, stores, and tracks electronic documents, i.e., word processing files, PDFs, digital images of paper-based content, and many more. The document management system saves time and money [5]. Apart from that, it gives some benefits like security of documents, centralized storage capacity, easy accessibility of control, streamlines search, audit trails, and retrieval. A document management system has the following essential features: (i) it has developed from digitized paper and ends with security, audit, and many more activities. Therefore less paper with having more functionality can be seen in this feature, (ii) document management system provides high storage capacity. In this case, a server's storage capacity is more significant than a physical filing cabinet, (iii) DMS has various scopes, but it has common attributes like indexing, version control, etc.

When the market is very competitive, large labor forces are highly skilled, and then it is evident that business firms are pretty concerned about document management systems. Paper storage requires large and significant physical space. This is a system that provides greater control, high accessibility, and effectiveness in the process. It gives so many benefits in the form of information retrieval, security, and governance of information. It is very cost-effective; the operational cost is marginal.

Moreover, proper and accurate record management acts as a legal imperative. The various versions of the document can be created and modified easily by the usage of DMS. This system has drastically changed the organization in the domain of reduction of paperwork [6, 7]. Digital document system is based on computer programs. A document management system is related to digital asset management, record management, document imaging and workflow management, and content management systems. The record management system is the process of managing records for the organization throughout the entire records-life cycle. This system includes systematic control of creating records, efficient maintenance and destruction of records, and business transactions associated with them. Through a digital asset management system, digital assets can be stored, shared, and organized in a central location in a business. It has a relation with creative files like images, videos, music, photos, and other media. These digital files are considered assets. The digital asset management or DAM is one of the systems connected with the company's content sharing and storage solution. Workflow management signifies the coordination of tasks in a business house. The term workflow can be described as a sequence of tasks coming under a more significant part of the task. Sometimes it is connected with the words 'business processes'. The objective of workflow is to accomplish some outcomes related to business activities; on the other hand, workflow management has its purpose, like achieving better results as per set goals [8].

An example can be given in this scenario, suppose when an organization hires new employees, there is some purpose, it has a clear business goal. A new employee who will join the organization has passed the stages of selection, and then, the person joins. For achieving this recruitment goal, recruiters, hiring managers, HR staff all work together on a sequence of tasks [9]. When the process of hiring goes, there is a need for several information such as the name of the candidate and resume, details of the job position, evaluation of individuals, interview time, candidate's performance, decision making to finalize the candidate, etc. This information should be captured, stored, and made available whenever required. Workflow management must be incorporated here.

2.2.1 Components of Document Management System

It is well established that a document management system is one of the essential tools under knowledge management. This platform provides information storage, metadata, versioning, indexing, security, and retrieval of information. There are various components of the document management system (Fig. 2.1).

The components are discussed below.

2.2.1.1 Metadata

Metadata has been stored for each document of various functionality. Metadata is such data that helps to recognize the document that the user stores. Let's take an example of the date on which the document will be saved or stored, and it gives the identification of the document. The metadata can be extracted with the help of DMS from the document automatically [10]. DMS also directs users to add metadata whenever it is required. The document management system can use optical character

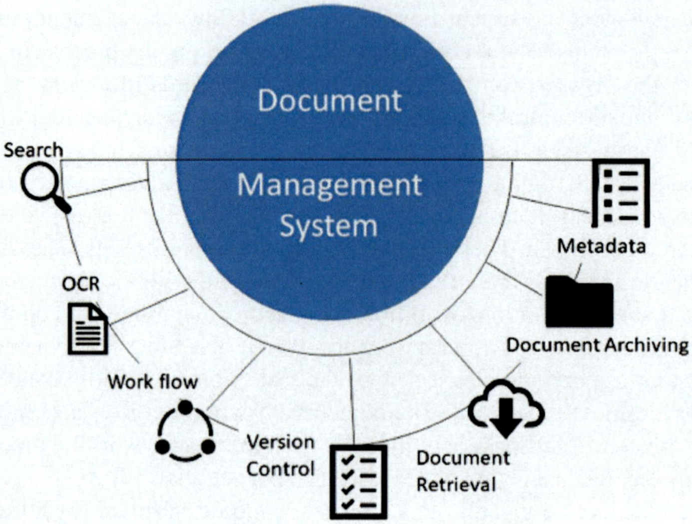

Fig. 2.1 Document management system

recognition on scanned images and text extraction on electronic documents. In this way, the extracted text is used to assist users to locate documents by recognizing probable keywords and provides search capacity for the full text. This extracted text is stored as one type of component of metadata. It is stored or saved with the document and can be separated from the document as a source for searching documents among other collections.

2.2.1.2 Integration

Many document management systems provide proper document management functionality to another type of application directly. Therefore the users can retrieve the existing documents directly from the document management system repository. It helps to make changes and save the changed or modified documents back to the repository [11, 12]. Here it acts as a new version without leaving the application. This type of integration can be available in various software tools like content management systems, workflow management. It is done with the help of an application programming interface or API. It uses open standards, i.e., WebDAV, SOAP, ODMA, LDAP, etc.

2.2.1.3 Data Validation

Data validation is also very much significant as data is generating information. Data validation monitors the accuracy of source data before using it, and it checks the

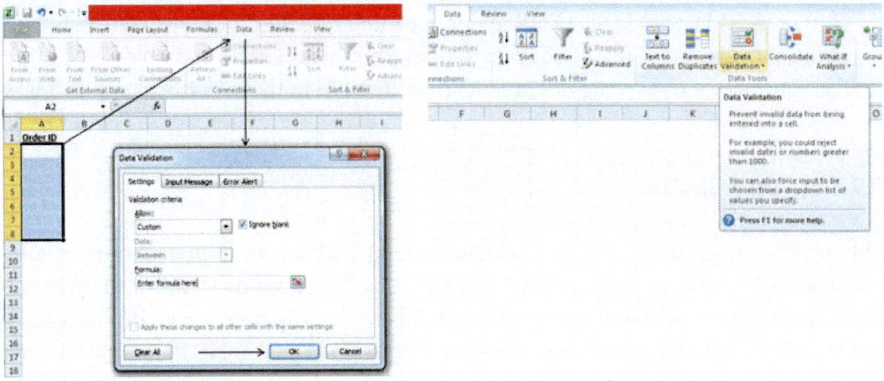

Fig. 2.2 Data validation in MS excel

quality of data. Based on objectives or destination constraints, different types of validation can be performed (Fig. 2.2). Data validation is a part of data cleansing. The rules of data validation are checking documents failures, names that are misspelled, missing signatures, etc. It recommends real-time correction options before data is imported into the document management system. In this part of data validation, a data format can change. Harmonization or additional processing of data are also included and applied in this stage. Data validation aims to make it consistent, accurate, and complete. In this way, data loss or errors during movement will be prevented. This provides benefits like shielding the transferable data from any corruption due to some inconsistencies in type or context.

2.2.1.4 Indexing

Indexing is the process by which electronic documents can be tracked. Indexing is a simple process of unique document identifiers. Sometimes appropriate indexing presents with more complexity when this process provides classification with the help of documents' metadata. Information query and retrieval are supported by indexing. A critical area for rapid retrieval is to create an index topology or scheme. A topology index is referred to as a connectivity index [13]. It is a type of molecular descriptor, and it is calculated based on a molecular graph. Topological indices are one type of the numerical parameter of a graph that helps to characterize its topology.

2.2.1.5 Capture

Capture, this feature is used for accepting and processing images of a paper document. This document comes from scanners or printers of multifunction. In this case, OCR or optical character recognition software has been used; it is integrated into the hardware. It acts as standalone software for converting digital images into machine-readable text [14]. OMR or optical mark recognition software has been used for extracting values of checkboxes and bubbles. Other computer-based files and acceptance of electronic documents may involve in the process of capturing.

2.2.1.6 Distribution

When a document is ready for distribution along with a format, it cannot be altered. The original or master copy of the document is not used for distribution anymore. In this case, an electronic link for the document is more common to use. In a regulatory environment, if a document is distributed electronically, it includes versioning and traceability assurance. The document can be interexchange by approaching DMS when document integrity is imperative.

2.2.1.7 Retrieval

Retrieval is another feature of a document management system (Fig. 2.3). The retrieval process takes place when there is a need to retrieve electronic documents from the storage. The process of retrieval in the electronic context is quite complex and powerful. On the other hand, the notion of particular document retrieval is straightforward. Unique document identifiers support the simple retrieval of individual documents by users. A flexible retrieval process allows the user to make specific partial search terms that involve document identifiers and few expected metadata parts. These return a list of documents that can be matched with the user's searching term. Data retrieval helps to identify and extract data from a database. It works when a query is provided by the user or at the time of using the application. It also enables data fetching from the database due to displaying it on a monitor. It can be used within an application as well. Figure 2.3 shows an image retrieval process.

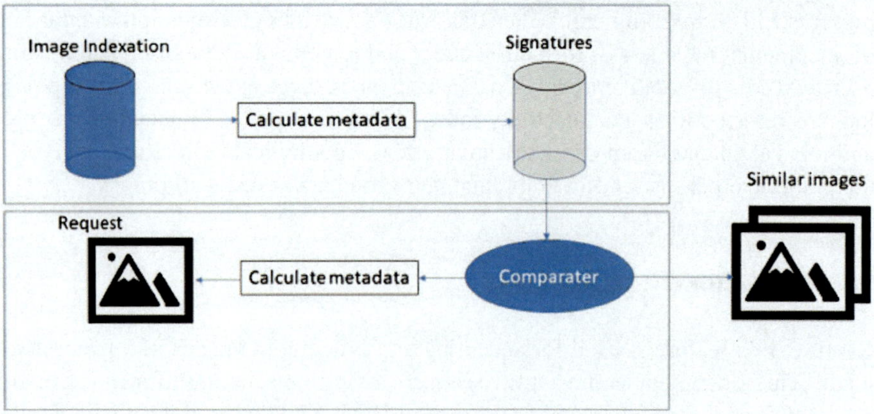

Fig. 2.3 Content-based image retrieval

2.3 Learning Management System

A learning management system (Fig. 2.4) is a particular type of software application. It provides a framework to handle all functions in a learning process. It identifies the training needs in a business and works accordingly. A learning management system has a relation with a training management system. Training is a process in the organization that employees can enhance their skills and knowledge for doing a particular work. The training process also increases the potentiality of employees for future work. The learning activity management system is also known as a learning experience platform.

On the other hand, learning a content management system is different from LMS. Learning content management system is another type of software that is used for managing content. The two systems are not the same. Instead, these two systems are complementary to one another [13, 15]. A learning management system has been designed to make life easy for those in charge of training and development functions in a business house. Learning management system identifies and assesses individual goals, as well as organizational goals. This system also tracks the path of progress toward the accomplishment of goals. LMS has other functions to play, like collecting and presenting data to supervise and monitor the process of learning. Besides these, LMS can handle and manage other business activities like onboarding, compliance, analysis of skill gap, and talent gap. It is a fact that most organizations

Fig. 2.4 Learning management system

are using SharePoint. It is a document management platform that enables browser-based collaboration. Microsoft SharePoint is a type of enterprise information portal. It has been configured for running intranet, extranet, and internet sites smoothly. SharePoint-based learning management system has streamlined the training process in a business environment. It allows the communication process to be more straightforward for the administrator. It efficiently manages content and collaboration. After that, training modules go into the LMS. Here everything is being housed under the same platform. It is also very user-friendly. Figure 2.4 depicts the different aspects of a learning management system.

2.3.1 *Benefits of Learning Management System*

Different benefits can be reaped from using learning management systems in organizations or institutions. Research has focused on innovations in both paradigms of diffusion as well as the adaption process. Business leaders use the learning management system, but it has also applied to the education sector. This helps to enhance the capability of a user in technical fields and provides success for organizations for the implementation of new technology. The advantages of LMS are as follows.

2.3.1.1 Cost Savings

Learning management system enhances the ability of training and teaching processes. Training is a process by which employees can increase their knowledge and skills for doing their present job. It reduces traveling of employees, optimizes the expenditures of training, and minimizes facilities to the investors [16]. From the aspect of cost-effectiveness, it is very much workable. In the educational field, learning management system is also used effectively to enable the relationship between teachers and students.

2.3.1.2 Consistency of Training

The learning management procedures are very consistent, as it is centralized. This delivers qualitative learning and teaching methodology to all employees and students in the education sector. Course materials, instructions, study notes, necessary documents, schedules, etc. can be uploaded through this system.

2.3.1.3 Easily Tracks Learner Progress and Performance

The learning management system helps to generate training reports based on users and students. Online training through e-Learning courses can be processed quickly

[17]. In this process, trainers can check and track the progress as per goals. This system has a positive relation with returns on investment and signifies the knowledge-gaining process.

2.3.1.4 Meet Regulatory Compliance

Many industries like oil and gas, communication, construction, pharmaceutical, etc. are training, accessing, and reporting information for compliance. A learning management system emphasizes making these legal and regulatory requirements. Apart from this, a learning management system has many more capabilities applicable to the business management field. From a business perspective, these are denoted as popular selling points, and top identifiers can judge the organizational effectiveness through this process.

2.4 Decision Support System

Under the era of tools on knowledge management, a decision support system is a basic computer program application. It is used to improve the decision-making capability of a company. It can analyze a large amount of data and produce the best solution. The DSS (Fig. 2.5) makes the best possible options for the organizations [18]. Data and knowledge can come from different areas and sources through a decision support system. This gives information to the user that is beyond the usual reports and summaries. It has an intention to inform people about decisions. A decision support system is one type of informational application; it is opposite to an operational application. Based on various data sources, informational applications

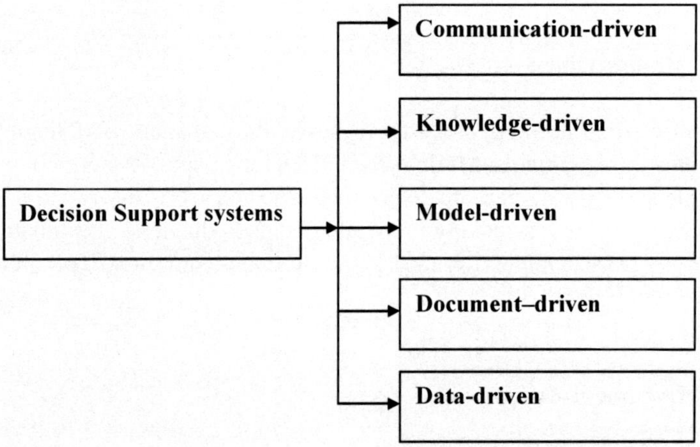

Fig. 2.5 Types of decision support systems

provide relevant information to the users. DSS supports a better decision-making policy in the organization.

On the other hand, an operational application records the details of business transactions, including data required for the decision support needs in a business house. There are few examples of a decision support system where it works effectively. For instance, a company wants to compare sales figures between a week and the following week. At that time, DSS helps to figure out the comparison and make the best possible outcome. If the company wants to map revenue figures (projected) based on new product sales, assumptions can be made through DSS [19]. The third case states that the consequences of different decisions also depend on a proper decision support system.

2.4.1 Types of Decision Support Systems

2.4.1.1 Communication-Driven

The communication-driven decision support system allows business firms to support tasks and roles where more than one person needs to work on that task. It includes integrated tools, namely, Google Docs and SharePoint workspace (Fig. 2.5).

2.4.1.2 Knowledge-Driven

The knowledge-driven decision support system provides concrete and specialized solutions to situations involving interactive decision-making structures, just like flowcharts. It also uses storing facts, procedures of work, and formation of rules and regulations.

2.4.1.3 Model-Driven

A model-driven decision support system accesses the management of financial, statistical, organizational models. Model-driven decision support system collects data to frame models, and parameters are determined based on users' information. The information creates a decision-making model that helps to diagnose the situations. Here, model-driven DSS includes Dicodess. It is an open-source model-driven decision support system [20].

2.4.1.4 Document–Driven

Document–driven decision support system helps manage unstructured and unorganized information into different types of electronic formats. Document–driven DSS integrates a variety of storage. This computerized system has processed technologies for providing document retrieval and then analysis. The system or subsystem has the intention to assist and support the decision-making process.

2.4.1.5 Data-Driven

The data-driven decision support system is used for storing and analyzing internal and external data of an organization. It has access to manipulate time-series data (internal and external) in a business. In this case retrieval tool has generated the most superficial level of functionality effectively.

2.4.2 Benefits of Decision Support System

A decision support system has the feature of increasing the speed and efficiency of the decision-making process. This activity is possible due to the collection and analysis of real-time data by DSS. It can promote training and development within an organization. Specific skills should be developed for the people of the organization. By this, they can implement and run a decision support system efficiently. Sometimes managerial processes become monotonous. It is because managers are spending most of the time on decision-making so that it takes so much time, and the process becomes lengthy. Managers cannot fulfill other tasks. Therefore; DSS provides the benefits to automate the decision-making process. Apart from that decision support system helps in improving interpersonal communication systems within the business house. A decision support system reduces time to make decisions in the workplace, examines the quality of decisions, and improves its effectiveness.

The business vendors sometimes cite these benefits with business intelligence systems, performance management systems, and web-based document management systems. This makes a competitive advantage for the firms. The vendor's regular sale of the product to clients in the market has become easy with the help of a computerized decision support system. The appropriate installation can also make cost reduction of DSS. Some case studies have shown that DSS has made low infrastructure costs and technology costs. A computerized decision-making system has reduced the investment in labor. The DSS helps to minimize the frustration of decision-makers and increase the level of satisfaction by the perception that individuals can make a better decision. The learning ability on new concepts and development of business environment can be increased by promoting training tools for new employees by using decision support system as well [21, 22]. Management can better understand various business operations with DSS. Managers perceive that a decision support system is very much needed and valuable in functional domains. Thus it increases the level of administrative control in the true sense of decision-making ability.

2.5 Social Communication Tools in Business

Social communication tools are very recognized in every business organization. Appropriate and effective communication is required in each sphere of management, whether small, medium, or large. In order to increase the efficiency level and productivity of the business, information technology management has given birth to user-friendly solutions. They focus on internal communication software to stimulate employees' motivation. Social communication tools in business capture knowledge, generate knowledge, and communicate knowledge. The various communication tools are speech recognition, video conferencing, and collaboration tools such as whiteboards, telephone services including VOIP, emails, chats, instant messaging, and many more. Unified communication tools have changed the integration in communication into a single user experience. The business communication tools address internal communication problems that organizations face and give solutions [23]. Some of the communication tools that are used in business for knowledge management purposes are briefed below.

2.5.1 Social Intranet Software

Every business organization has a high requirement for practical social communication tools. The approaches regarding this issue are implemented with the help of information technology to make the business go forward. Social intranet software helps to lower down IT maintenance costs and save business money. By using telecommuting and BYOD, an intranet solution can cultivate a more flexible workforce in the workplace. It has increased productivity to align all employees with the same goal. Social intranet creates a communication and collaboration hub that can be easily accessible to get the work done. This is a user-friendly tool incorporate. It provides benefits for the creation of multimedia and powerful searching tool. As per statistics, it is said that more than 75% of people are using their intranet solely for HR purposes.

2.5.2 Private, Group Messaging, and Chat Tools

There are different communication tools, i.e., private messaging, group messaging, chat tools in the business. By using these communication tools, knowledge can be shared. The project teams need collaboration tools for making the work more effective. To be informed all time, the team members need proper communication with each other. Without having a platform of communication, it might create a problem of collaboration, coordination, and cooperation. Team members face difficulties when there is a need to share ideas and information. Sometimes emails can be lost or can remain unread. Different chat tools, private or group messaging, etc., have played a vital role in the team's progress and productivity.

2.5.3 *Task Management*

Business organizations use task management tools for completing project tasks on time. It can be used by an individual or team or an entire organization as well. The tool helps to organize and prioritize the related tasks of a project. There are various task management tools, such as basic spreadsheets, different project management applications online, etc. Task management tool makes the business performance so much efficient. If there is lack of task management tools, the performance of the business will not be successful. These tools efficiently help the workforce to complete their tasks in a most vital aspect. The jobs become meaningful and essential. When task management tools combine with robust communication features, it raises the tracking capabilities of the assignment. In this scenario, an example can be given that the project may face difficulties if a project management system is absent. Absence of project management system will lead to prolonged time for completion of project by team members. These issues can be resolved when management escalates an effective task management tool to monitor progress [24].

2.5.4 *Internal Blogs and Videos*

Another crucial social communication tool in business is internet blogs and videos. Content is such a thing that defines whether internal teams are engaged or not at their work. Content can give a clear idea about the organization and its people and their duties. If the content looks great, people will read more and more. Through these blogs and videos, people can be highly motivated and engaged in their tasks. They become informed about the roles that lead to more excellent performance. It makes participants active if the type of content is attractive.

2.5.5 *Employee Profiles and Workflows*

Every business needs to know about their staff. Every staff needs to know each other. It does not matter whether the company structure is tiny, medium, or large. Relatively strong collaboration is needed at every level of management. Therefore, people and their roles within the organization are critical, and communication plays a vital role. Every employee profile should be known to make communication easy, and others must understand their duties in the organization. Workflow management is also playing a prime role in this field [25, 26]. It has been observed that sometimes project managers are very busy with different tasks, and therefore the team members do not get the correct information at the right time. It may create a delay in the completion of the project. Projects suffer from late approval also. Teams can face some uncomfortable positions due to this delay. Therefore proper workflow management is significant in the workplace for connecting and sharing knowledge at the right time at the right place.

2.6 Customer Relationship Management System

A customer relationship management system helps the functional domains like sales, marketing, customer service, etc. (Fig. 2.6). The marketing team tracks prospect and gather customer information through this system. With the help of a customer relationship management system, it is monitored how customers are getting informed about the company or product and what type of information they have earned from the website. Then over the buying cycle, the sales team can interact with that customer and track it properly. The interaction will be continued after that in the form of repeated purchases. A knowledge management tool is like a customer relationship management system [27]. It enables the entire team to understand customers better and also maintain a good relationship. The response is very much positive and continuous from the firm's side toward the prospects. The CRM system is a particular type of technology that helps manage all kinds of relationships of a company, such as

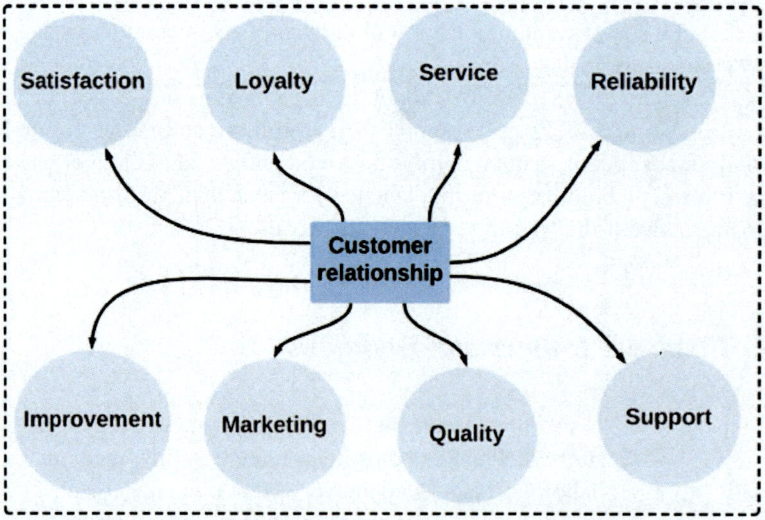

Fig. 2.6 Customer relationship

client-based, customer-based, investor-based, and so on. This emphasizes interaction with customers and potential customers as well. The objective of this system is straightforward that is to improve business relationships. Customer relationship management system always focuses on streamlining processes, staying connected with customers, and increasing the profitability of the business. Nowadays, CRM is mainly referred to as customer relationship management software. It is a tool that acts as a single repository system [28, 29]. The company's sales, marketing, customer service activities all are coming together under CRM software. This also helps to streamline the business process, policy, procedures, and people in a single platform. Remote CRM is a process that makes the remote sales team productive and provides team members necessary tools for collaboration with their peers and customers.

2.6.1 Features of Customer Relationship Management System

CRM software has become the world's most powerful software. It has proved that it is the best technological asset in a company. Businesses must invest in this software. A customer relationship management system helps the implementers to make every aspect of the business cycle effective. CRM software can easily integrate with other used applications in a business. It gives a hike in marketing and sales returns. There are some essential features of CRM as follows.

2.6.1.1 Lead Management

A customer relationship management system can track and monitor the leads, and it is very much focused on converting leads into happy and paying customers. Leads can be generated based on a variety of parameters. After analyzing, the decision will be taken on which lead has the highest chance to convert into the pipeline.

2.6.1.2 Contact Management

The feature of contact management helps to maintain contacts of a business firm in one place. By using a customer relationship management system, a firm can quickly know when and where the customer has been contacted by the firm last time and what will be a suitable time to reach them for interaction after that. In this case, emails and other forms of social communication can be made [30].

2.6.1.3 Sales Automation

Naturally, the sales team should be engaged most of the time in selling activity rather than other administrative tasks. A CRM system has done the automation of the sales process in the workplace. This system makes the sales automation quickly where the manual process is the time taken. Leads will come automatically into the business pipeline with the right sales reps.

2.6.1.4 Marketing Automation

Due to poor planning and faulty target audience, sometimes companies face wastage of marketing budget for their brand. If there is no coordination between sales and marketing teams, this problem can generate in the workplace. Marketing automation and campaign management can build quickly by a customer relationship management system (Fig. 2.6). The marketing automation features help generate new leads, compare sales revenue, calculate expenditure, execute a target marketing campaign, and estimate the maximum return on investment on marketing spend.

2.6.2 Benefits of Customer Relationship System

The customer relationship management system is helpful for salespersons, marketing teams, and customer service providers or professionals. It applies to the people of the organization who have direct or indirect interactions with customers. The prime benefits of the CRM system are discussed in this section.

2.6.2.1 Keeping Track of Tasks

The software of CRM has reduced the amount of time spent on routine jobs and maintained a standard work process. Marketing automation has increased customer engagement and consistency [31]. One of the hidden advantages of the CRM system is consistency. If the customers receive conflicting signals, it makes them puzzled; enthusiasm will be lost. To eliminate this problem, a customer relationship management system keeps track of tasks and activities.

2.6.2.2 Increased Sales Prospects

Companies sometimes face difficulties expanding their client base and cannot make innovative ways of lead generation. A CRM system is helpful to learn about existing clients. This system tracks the purchase patterns of the customers and discovers new

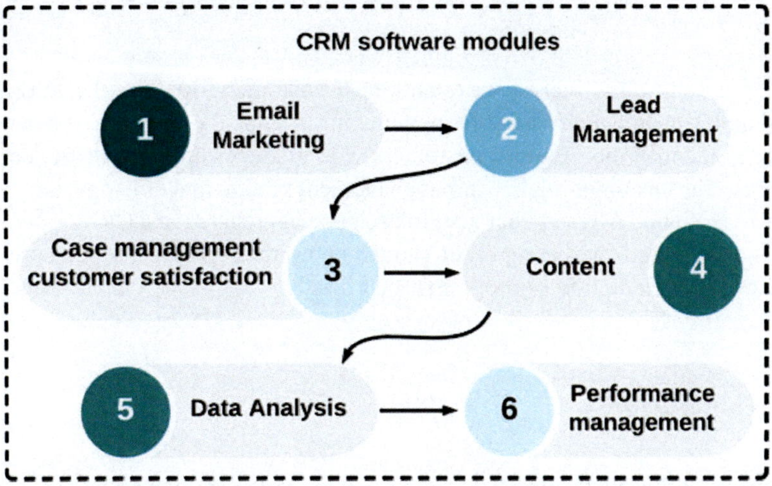

Fig. 2.7 CRM software modules

trends and themes as well. In this case, collecting data is creating micro-targeted strategic plans to enhance sales for a certain period.

2.6.2.3 Keeping Track of the Company's Sales Progress

Decent CRM software (Fig. 2.7) is quite powerful to make a business efficient and effective. It measures sales activity for the entire company along with client base management. Whenever there is a requirement of tracking leads and follow-up, the customer relationship management tool helps the business a lot [32, 33]. If any person or sales agent has an extensive sales force, the system can help him or her to monitor the path of work that is easy to access later. Each fresh set of leads must be allotted to the sales agent by the company. Therefore it is their need to have a proper CRM system. A CRM system helps to know the sales trends and closing gaps in the journey. Overall sales performance can be inspected and judged by the effective installation of a customer relationship system in the workplace.

2.6.2.4 Individualize the Communication

Many CRM solutions offer individualization in communication. A study has revealed that more than 86 percent of consumers believe personalization is very effective in making their purchase decision. For customer interactions, all of the information of the CRM database is utilized. It also develops close customer connections. Their first names in the emails can identify customers. In this way making individual communication becomes effective for higher sales.

2.6.2.5 Increased Retention

A CRM system also focuses on retention of customers. It gives the information regarding most engaged or most money-making clients. By using CRM software, it is easy to identify the group of customers who read all company emails or attend webinars. The customer relationship management system makes happy and satisfies the current clients. All customer activities can be aggregated in a CRM system [34–37]. The sales and marketing team should know their prospect's perception and behavior to develop plans or programs accordingly.

2.7 Conclusions

Knowledge is the most important and valuable asset for all individuals who are working in different industries. Knowledge management applies to small-sized, medium-sized, and large-sized companies. With the help of knowledge management tools, unnecessary faults can be minimized in business functions. It leads to high and qualitative performance of the business. Managers are people who have lots of duties in the workplace with more significant variation. Based on existing data, information, and knowledge, the managers and staff can use their skills to analyze, understand the problems and lead the best way to achieve the goals. It is the duty of the senior managers that they should make the work environment comfortable and better. Knowledge should be generated and shared by all people in the workplace. Knowledge management tools help to complete the process quickly. Therefore knowledge management tools are very much significant for all levels of management. Knowledge is not perishable; it should be stored and regenerated with new things. Employees need to motivate all the time to update their knowledge and learn something new every day. Firms are competitive. To build a positive and attractive image in the market, every firm should practice AI-driven knowledge management. Organizational performance measures the success of business units. If knowledge is managed correctly by various authentic tools, then the process of knowledge management will have an impact on the developmental factors of an organization. Strategic management will be meaningful in the practical field. Sharing insights and experiences is indeed very much crucial to improve programming. This is the central aspect of knowledge management. KM tools have critical roles in every function of the business. Sometimes it can be said that the knowledge management process is the heart of the company. With the development of information technology, knowledge management tools help the users perform their work better and maintain it effectively. The company has a goal to accomplish. The KM tools constantly analyze the strength and weaknesses of the companies and try to find the best direction to solve them. In this chapter, we have discussed various types of knowledge management tools. All these have an individual influence on business. It emphasizes smooth running of the firms and, finally, accomplishing the objectives.

References

1. Ruggles, R. (2009). *Knowledge management tools.* Routledge.
2. Rao, M. (2012). *Knowledge management tools and techniques.* Routledge.
3. Ghani, S. R. (2009). Knowledge management: Tools and techniques. *DESIDOC Journal of Library and Information Technology, 29*(6), 33.
4. Ngai, E. W., & Chan, E. W. C. (2005). Evaluation of knowledge management tools using AHP. *Expert Systems with Applications, 29*(4), 889–899.
5. Tang, A., Avgeriou, P., Jansen, A., Capilla, R., & Babar, M. A. (2010). A comparative study of architecture knowledge management tools. *Journal of Systems and Software, 83*(3), 352–370.
6. Oliva, F. L., & Kotabe, M. (2019). Barriers, practices, methods and knowledge management tools in startups. *Journal of Knowledge Management.*
7. Massingham, P. (2014). An evaluation of knowledge management tools: Part 1—Managing knowledge resources. *Journal of Knowledge Management.*
8. Hoffmann, I. (2001). Knowledge management tools. In *Knowledge management* (pp. 74–94). Springer.
9. Balmisse, G., Meingan, D., & Passerini, K. (2007). Technology trends in knowledge management tools. *International Journal of Knowledge Management (IJKM), 3*(2), 118–131.
10. Massingham, P. (2014). An evaluation of knowledge management tools: Part 2—Managing knowledge flows and enablers. *Journal of Knowledge Management.*
11. Vaccaro, A., Parente, R., & Veloso, F. M. (2010). Knowledge management tools, inter-organizational relationships, innovation and firm performance. *Technological Forecasting and Social Change, 77*(7), 1076–1089.
12. Alwert, K., & Hoffmann, I. (2003). Knowledge management tools. In *Knowledge management* (pp. 114–150). Springer.
13. Grimaldi, M., & Rippa, P. (2011). An AHP-based framework for selecting knowledge management tools to sustain innovation process. *Knowledge and Process Management, 18*(1), 45–55.
14. Verran, H., Christie, M., Anbins-King, B., Van Weeren, T., & Yunupingu, W. (2007). Designing digital knowledge management tools with Aboriginal Australians. *Digital Creativity, 18*(3), 129–142.
15. Prokopiadou, G., Papatheodorou, C., & Moschopoulos, D. (2004). Integrating knowledge management tools for government information. *Government Information Quarterly, 21*(2), 170–198.
16. Spector, J. M. (2002). Knowledge management tools for instructional design. *Educational Technology Research and Development, 50*(4), 37–46.
17. Heathfield, H., & Louw, G. (1999). New challenges for clinical informatics: Knowledge management tools. *Health Informatics Journal, 5*(2), 67–73.
18. Pfeffer, K., Baud, I., Denis, E., Scott, D., & Sydenstricker-Neto, J. (2013). Participatory spatial knowledge management tools: Empowerment and upscaling or exclusion? *Information, Communication and Society, 16*(2), 258–285.
19. Duffy, J. (2001). The tools and technologies needed for knowledge management. *Information Management, 35*(1), 64.
20. Jackson, C. (2001). Process to product: Creating tools for knowledge management. In *Knowledge management and business model innovation* (pp. 402–413). IGI Global.
21. Ruggles, R. (1998). The state of the notion: Knowledge management in practice. *California Management Review, 40*(3), 80–89.
22. Dieng, R., Corby, O., Giboin, A., & Ribiere, M. (1999). Methods and tools for corporate knowledge management. *International Journal of Human-Computer Studies, 51*(3), 567–598.
23. Jenab, K., & Sarfaraz, A. R. (2012). A fuzzy graph-based model for selecting knowledge management tools in innovation processes. *International Journal of Enterprise Information Systems (IJEIS), 8*(1), 1–16.

24. Razmerita, L., Kirchner, K., & Sudzina, F. (2009). Personal knowledge management: The role of Web 2.0 tools for managing knowledge at individual and organisational levels. In *Online information review*.
25. Alberghini, E., Cricelli, L., & Grimaldi, M. (2010, March). Implementing knowledge management through IT opportunities: Definition of a theoretical model based on tools and processes classification. In *Proceedings of the European Conference on Intellectual Capital* (pp. 21–33).
26. Dalkir, K. (2013). *Knowledge management in theory and practice*. Routledge.
27. Nazim, M., & Mukherjee, B. (2016). *Knowledge management in libraries: Concepts, tools and approaches*. Chandos Publishing.
28. Cupiał, M., Szeląg-Sikora, A., Sikora, J., Rorat, J., & Niemiec, M. (2018). Information technology tools in corporate knowledge management. *Ekonomia i Prawo. Economics and Law, 17*(1), 5–15.
29. Varun Grover, T. H. D. (2001). General perspectives on knowledge management: Fostering a research agenda. *Journal of Management Information Systems, 18*(1), 5–21.
30. Cobos, R., Esquivel, J., & Alamán, X. (2002). It tools for knowledge management: A study of the current situation. *Journal of Novática and Informatik/Informatique, 3*(1).
31. Pettenati, M. C., & Ranieri, M. (2006). Informal learning theories and tools to support knowledge management in distributed CoPs. In *Proceedings of Innovative Approaches for Learning and Knowledge Sharing* (pp. 345–355).
32. Lai, Y. L., & Lin, F. J. (2012). The effects of knowledge management and technology innovation on new product development performance an empirical study of Taiwanese machine tools industry. *Procedia-Social and Behavioral Sciences, 40*, 157–164.
33. Youn, S., & McLeod, D. (2006). Ontology development tools for ontology-based knowledge management. In *Encyclopedia of E-commerce, E-government, and mobile commerce* (pp. 858–864). IGI Global.
34. Talamo, C. (2015). *Knowledge management and information tools for building maintenance and facility management*.
35. Karaa, W. B. A., & Dey, N. (2017). *Mining multimedia documents*. CRC Press.
36. Dey, N., Ashour, A. S., & Nguyen, G. N. (2020). *Recent advancement in multimedia content using deep learning*.
37. Bureš, V., Tučník, P., Mikulecký, P., Mls, K., & Blecha, P. (2016). Application of ambient intelligence in educational institutions: Visions and architectures. *International Journal of Ambient Computing and Intelligence (IJACI), 7*(1), 94–120.

Chapter 3
Knowledge Management in Various Sectors

3.1 Introduction

The most important factor of production is knowledge. It is next to labor, land, and capital. For development of the organization, managing and sharing knowledge is very much essential. In the modern competitive environment of business, knowledge management has played a pivotal role in sustainable development of organizations. In the twenty-first century, knowledge and knowledge management both act as professional elements in various fields of knowledge; namely, health care, education, construction, sociology, management science, psychology, philosophy, economics, knowledge engineering, artificial intelligence, information and technology, and other branches of business [1]. With the help of successful implementation of knowledge management, all business firms, irrespective of size, can improve their effectiveness. KM provides valuable ideas in decision-making process for the benefit of a company. The application of knowledge management leads to higher level of efficiency. Knowledge Management with AI has made better performance of the organizations in the market; enhanced capabilities of new manpower and qualitative decision-making procedures [2]. Keeping in mind the extreme importance of Knowledge Management in businesses, we have discussed KM practices in different sectors in this chapter, mostly focusing on health care, construction, education, SMEs, and E-business. In Sects. 3.1.1 Knowledge Management in health care industry, in 3.2.1 Knowledge Management in construction sector, in 3.3.1 Knowledge Management in education sector, in 3.4.1 Knowledge Management in SMEs and finally, in 3.5.1 Knowledge Management in e-business are reported. Section 3.6 is conclusion section.

3.1.1 Knowledge Management in Health Care Industry

Knowledge management is essential to understand the performance of an organization. Recent studies have revealed knowledge management has a positive impact

S. Majumder and N. Dey, *AI-empowered Knowledge Management*,
Studies in Big Data 107, https://doi.org/10.1007/978-981-19-0316-8_3

on operations and overall performance of the organization. Health care delivery is a knowledge-driven process for the improvement in all levels of processes through knowledge management capacity. Some recent reviews on knowledge management have stated the information that it gives an important insight into the health care industry. Three special domains of KM like knowledge creation, knowledge normalization, and knowledge application have complicated the KM process in health care industry. Personal values, norms of society and objective facts are influenced by these three domains of knowledge management. Besides these, there are three major themes that are very important in health care, namely, (i) nature or characteristics of knowledge in health care, (ii) types of knowledge management tools and their initiatives, and (iii) barriers in the process of KM. Basically, the paradigm of knowledge management in health care is very much new [3, 4].

3.1.2 Healthcare Delivery and Performance in KM Context

The health care providers maintain an inter-organizational relationship with other players in a foundation. Delivery of health care is very complex in nature. Health care organizations vary from other organizations on the basis of various aspects. In case of health care, the nature of the work is variable and complex; it might be an emergency or of non-deferrable nature. In this service industry, there is very little tolerance for ambiguity or error. Accessibility and quality of health care delivery can come under huge pressure in case cost of health care delivery is high. Furthermore, accountability is also increased due to high rate of medical errors and due to globalization, high standard quality is desired. In this way, a very unique situation exists in the area of health care delivery. Primary loyalty belongs to their profession rather than focusing on the organization. The organizational performance is multidimensional in nature and has different functional streams like human resources, finance, training and learning, sales and marketing, etc. In every functional stream, there is an effect of knowledge management process in the context of health care sector. Through different domains of activities, performance in health care delivery can be studied and evaluated by its effectiveness and efficiency. Knowledge management tools are applied in various activities of health care delivery [5]. Clinical arrangement and patient care should be viewed in terms of outcome and degree of expectations of patients. Technical and interpersonal care plays a significant role in achieving the same. Interpersonal care has been defined as the psychological and social interactions between patients and physicians or other caregiver persons. In interpersonal care, customer support tools and document support tools of knowledge management are very much crucial. Under conceptual framework of knowledge management, it is noted that effectiveness of knowledge management system can be measured in healthcare delivery through impact on organizational level and patient care level. The evaluation process continues and is based on knowledge management infrastructure, technology, and knowledge management process capability including knowledge acquisition, knowledge conversion, knowledge application, and knowledge protection [6]. Finding, sharing, and developing knowledge leads to better decision making and organizational learning, which includes performance of entire organization in relation to quality, satisfaction, and production capacity.

3.1.3 Social Practices of Knowledge Acquisition and Sharing

Research revealed that large amount of scientific data and information of health care sector cannot be kept or tracked by individual health care professionals. Across the organizations and professional boundaries, the nature of knowledge in health care industry is disseminated and fragmented. Health care providers are overwhelmed with vast information but are not able to locate or isolate any particular information when required. In health care delivery, different professional groups have different rules and regulations, representation of jobs, behavior, and values in a process of inter-collaboration [7]. Here, explicit and tacit both kinds of knowledge aspects are incorporated for achieving outcomes in relation to quality, accessibility, and cost-effectiveness. Networking health care professionals are not a new thing; they have always played a prime role in health care delivery. On the other hand, formalization of these networks and sharing knowledge through this is very much new. Knowledge management system helps to share clinical evidence between different specialized health care professionals such as technicians, nurses, physicians, non-technical staff, etc. A community of practice has provided a mechanism for existing professional groups within health care organizations and to make leverage on their tacit knowledge base.

3.1.4 Knowledge Acquisition and Sharing Through Electronic Medical Records

Nowadays the traditional relationship between physicians and patients is moving in a direction that concerns a teamwork among healthcare professionals. Electronic medical records basically denote a patient's medical reports, they can be stored and retrieved in electronic or digital format. The objective of medical records should act as repository of physicians' observations and patient analysis. Medical record is full of historical information related to the patient, essential information on physical examination, test results, and procedures that are performed on the patient. Electronic medical record or EMR is defined as a process that enables storage of healthcare information throughout an individual's entire life with the purpose of continuous care, education, research, and development and to ensure confidentiality at all times. Data must be entered into the EMR directly by physicians or any other healthcare service providers. It can include a series of different techniques. The implementation and usage of EMR have an impact on the process of health care delivery in multiple ways, such as the way of interaction between physicians and nurses with patients, waiting time involved in admission and discharge of patients, etc. If longer EMR process is implemented in health care settings then as per analysis the patient satisfaction level decreases [8–10].

3.1.5 Knowledge Assimilation and Application Through Clinical Decision Support Systems

A patient-specific assessment or recommendation can be made with the help of clinical decision support system. The patient-specific information is analyzed and compared with an expert knowledge base and thus CDSS generates clinical decision-making ability. There are so many benefits of CDSS among health care providers. They can easily access clinical guidelines, drug dosage, prescription, computed aided diagnosis, electronic reminders, etc. With the help of CDSS, effective data mining has become one of the prime tools to acquire and operate health sector knowledge. Thus, explicit and tacit knowledge gets a chance to distribute among professionals [11]. As there is a revolutionary increase of clinical knowledge, health care professionals face dilemma to utilize health care processes. Generation of electronic repository in this area has made the management of tacit and explicit knowledge of health care professionals easier in the form of articles, guidelines, clinical protocols, etc. Clinical decision support system makes a synergy between data mining or procurement of knowledge and usage of knowledge in decision making by its functional and architectural specifications. CDSS has few important features like (i) it provides support in decision making automatically as a part of clinical workflow, (ii) it delivers decision support to professionals and location of decision making, (iii) it generates recommendations including actions. CDSS provides data for taking administrative and operational level decisions. This data is stored in data warehouses initially, to be used later d for the purpose of cost evaluation, performance evaluation, and review. The practice of evidence-based medicine is facilitated by particular CDSS in knowledge management practices. In collaboration with electronic repositories, CDSS provides the framework for achieving evidence-based healthcare [12]. Furthermore, CDSS helps to make triple assessments in breast cancer care and also establishes the logic of using this system, thus results may improve. Finally, the statement can be established that clinical decision support system has a positive relation with evidence-based medicine.

Health care industry has four types of contingency factors of clinical decision support system which gives impact on KM practices. These factors are denoted as characteristics of physician, ailment characteristics, organizational infrastructure on IT, and organizational characteristics. The clinical problem-solving paradigm is affected by physicians' technical skills, level of experience, and training. Combination of these factors will lead the healthcare delivery organization to make their choices on adaption of appropriate KM tools and to be highly focused on performance measures. Implementation of knowledge management tools and techniques in health care organizations is very much dependent on human resource management practices, nature of leadership, organizational culture, and infrastructure of IT. At this juncture, it can be said that appropriate integration of IT systems is a crucial aspect. Various committees are involved in the context of hospital settings and knowledge is a prime factor that needs to be shared among hospitals' various patient care teams (PCT). This PCT team is responsible to revise and draft evidence-based clinical guidelines or protocols and also conducting risk analysis. It is important to transmit knowledge among patient care team members [13, 14]. Lastly, we can mention acquisition of knowledge and it is sharing under healthcare delivery organizations are made and maintained by several appropriate KM tools.

3.1.6 Knowledge Management in Construction Sector

The construction industry is basically a project-based industry. Each of the projects is unique in nature and gathers a large number of stakeholders; they have collaborated with each other at various levels during a project life cycle [15]. Each of the construction projects has been considered as a unique multidiscipline organization that may be or may not be continued to be working together when the project is handed over. A significant complex process is generated due to heavy fragmentation of construction industry. It is an information-intensive industry where stakeholders communicate large-scale information across the various levels of project lifecycle. It is very much different and difficult to combine information management and knowledge management in this industry. If it is not managed properly it will create poor efficiency in the entire process. The industry is very much focused on improvement of construction processes. Pressure is also a part of these processes as clients' demands are high for better products within a shorter duration by utilizing fewer resources. In this scenario, keeping in mind the importance of information, construction industry has generated effective knowledge management as a strategy to promote innovation and improve the construction processes. One debate is continuing in the construction industry for long, i.e., this industry is very slow toward changes or adapting innovative solutions which are established by Information and Communication Technology sector or ICT sector. The other industries like automobile, manufacturing, etc. have successfully utilized ICT benefits for improving efficiency of its processes. With the changing phase and demand in the last decades, the construction industry has made outstanding efforts by adapting to ICT solutions. Figure 3.1 shows knowledge management process in construction projects. In this figure, knowledge acquisition has been segmented in two ways, namely s explicit knowledge and tacit knowledge. Explicit knowledge includes contracts, reports, drawings, general documents. Whereas, tacit knowledge includes process record, problem-faced, problem solutions, expert suggestions, innovation, know-how, etc. Both explicit and tacit knowledge creates best practices on knowledge database. Finally, this leads to knowledge-based decision support.

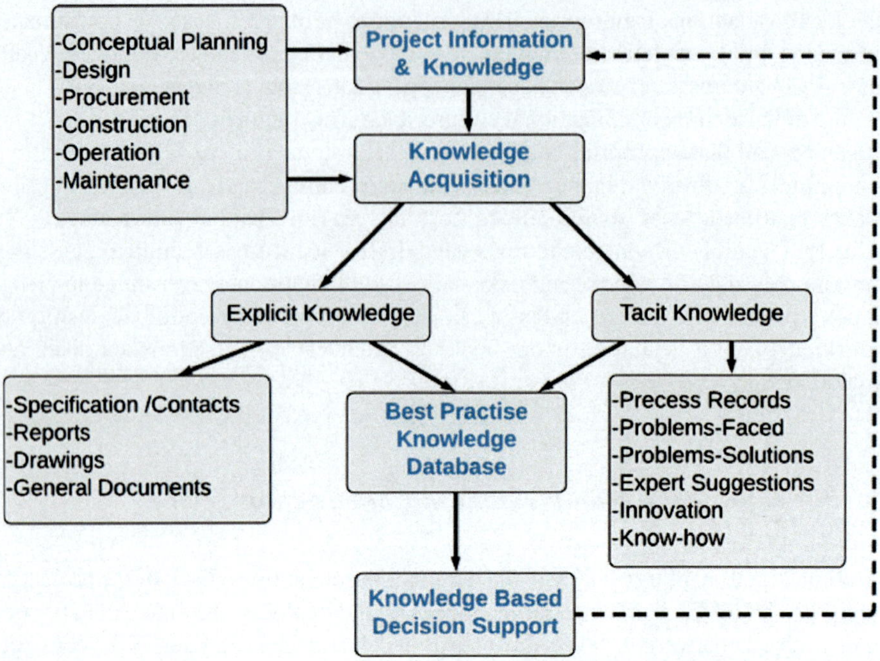

Fig. 3.1 Knowledge management process in construction project

3.1.7 A Notion of Knowledge Management in Construction

Each stage of a project life cycle represents a number of potential opportunities for capturing knowledge in the construction industry. Construction projects are unique and dynamic in nature. It is realized by the industry that if knowledge can somehow be captured and reused, it can reduce the waste and improve the efficiency of the process in general. As the industry is heavily fragmented, therefore capturing and sharing knowledge via common platform is difficult [16–18]. The outcome is that valuable knowledge is being lost. The industry involves a huge database of explicit knowledge. For gaining an advantage in the market, this industry must capture and reuse tacit knowledge. As per previous literatures, it is a big challenge to capture tacit knowledge. Within the construction industry, the current processes and systems do not support the capture of such tacit knowledge. It is important to mention in this area the emergence of technologies, i.e., Web 2.0 which provides an opportunity by applying innovative ways to create and capture tacit knowledge. It can be done through social interactions over the internet. Some key benefits of knowledge management in the construction industry are discussed below.

3.1.7.1 Enable Knowledge Transfer

Construction projects have a high requirement of knowledge. It flows through various stages like knowledge capturing to make design, estimation of actual construction, and so on. As this industry suffers due to heavy fragmentation, it is a very important aspect of knowledge management in construction industry. In most of the projects, there are various responsibilities toward the tasks. This helps toward smooth flow of knowledge across stages of the project. As a result, few problems arise like disputes on contracts, extensive rework, overrun of time and cost, etc. Under the construction industry, knowledge management has played a key role in facilitating effective knowledge transfer across various different levels of construction projects [19].

3.1.7.2 Capture and Reuse Project Knowledge

In this industry, there is a general consensus that the industry is not able to retain project knowledge for future reuse purposes. Few common particulars are behind this, namely, teams are separated after finishing the project, personnel change companies or industry over the time, lack of standard category platform for capturing and sharing knowledge, lack of motivating factors, etc. [20]. Many companies follow documentation processes like post-project reviews. Sometimes it is found that knowledge is not properly documented and even if documented, then they are locked in archives. Therefore, there is a need for introducing knowledge management system that helps to capture and reuse knowledge during and after finishing project tasks. For making best practices, such systems help to build database also.

3.1.7.3 Enable Better Communication Among Stakeholders

The construction industry is suffering from weak communication among stakeholders due to high fragmentation. This is one of the most prime aspects for transferring knowledge in this industry as many projects under construction run with problems such as cost and time overrun, lack of communication, disputed contracts, and many more. An effective knowledge management system always helps to communicate among stakeholders and preserve knowledge efficiently across project's various stages. Even the benefits of knowledge management are well documented through KM system. Nowadays, the construction companies are very much focused on collaborative knowledge and management solutions.

3.1.8 Social Media Platforms for Knowledge Management in Construction

In the construction industry, the building blocks of a social media platform in a framework include identity, relationships, presence, groups, reputations, conversation, and sharing knowledge. It is not essential that all social media platforms exist with all these elements. Focus is on some elements of these blocks. Many organizations have implemented Web 2.0 technologies in the forms of blogs, wikis, web-based forums, and social networks for capturing and sharing knowledge. There are few benefits of Web 2.0 over traditional knowledge management technologies, they are (i) easy to use, (ii) searching capabilities, (iii) availability of open-source, (iv) potentiality to re-create a virtual social environment, and (v) promotion of discussions. These kinds of social platforms have high potentiality to capture tacit and explicit knowledge [21]. This knowledge is created and shared every day in projects among different stakeholders. Cross-project learning is the final outcome. Previous litterateurs suggest that community model with personalization strategy is better than database approach. Later some arguments have been made by researchers that usage of social web-based tools may be regarded as complementary to facilitate capturing and sharing of tacit knowledge. They identified few main factors that have an impact on employees' knowledge sharing behavior in the workplace, based on social media platforms. These include personal, organizational, and technological factors. Personal factors like benefits of social media and employees' experience, organizational factors such as colleagues' activities, collaboration features, guidelines, and technological factors such as user-friendliness, requirement of skills, etc. are incorporated in this section. Besides these, there are some prime requirements in a social media platform that facilitates tacit knowledge. It includes social interaction, observation, sharing experience, informal relationship or networking, mutual trust, and many more. In other industries like information technology, manufacturing, oil and gas, automobiles have accessibility of usage of social collaboration platforms such as IBM connections, SharePoint, confluence, etc. for capturing and sharing knowledge. Due to temporary nature of construction projects, heavy fragmentation, resistance to change, etc. it becomes a challenge for application of such platforms in construction industry [22]. Furthermore, it is needed to mention that adaption process of ICT development in construction industry is very slow. Though the industry is able to capture explicit knowledge that is generated on projects there is no initiative to develop systems for capturing tacit knowledge. By the use of social media platforms, the industry is trying to capture and reuse project knowledge, maintain accuracy of captured knowledge, knowledge facilitation, avoidance of legal issues, make additional costs for workers, and work overload. Capturing knowledge and reusing them can be possible by web-based knowledge base like integrated workflow system. This system has some benefits, i.e., rework reduction, tacit knowledge sharing and retaining, and encouragement on innovation, client satisfaction, continuous improvement, and learning organization. In this industry, a mobile lesson–learning system application is allowed to create, evaluate and search lessons inside the database. Users have options to post micro blogs that are similar to twitter. It can direct to a particular person or group of people. The lessons which are learned must be collected in a structured format. Here users are asked to enter fields, namely, lesson title, description, date of creation, classifications, and name of the lesson's approver. In this way, social media platform has a great significance in the area of knowledge capturing and sharing in the construction industry. Sometimes this work industry follows BIM-based (BIM is a new platform for information technology applied to construction and implemented in new tools) social platform for effective knowledge management [23, 24].

3.1.9 Organizational Implication in Knowledge Management

Nowadays, it has been very much crucial to examine firm's performance. This may create problems in case of clients' dissatisfaction, low margin of profit-making, over capacity of resource but not utilized as optimum level. Senior managers of construction industry, therefore, are becoming more aware of different management principles and philosophies on holistic approaches to performance. Knowledge management has played a great role in the area of key performance indicators. Here we can take the example of Balanced Scorecard Model set by Kaplan and Norton. Learning and growth of the organization, internal business affectivity can be evaluated with the help of Balanced Score Card. For enhancement of project performance, the industry has emphasized KPIs or key performance indicators and made best practice programs.

3.1.9.1 Need for Organizational Strategy

Organization strategies are formulated for decision-making purposes without establishment of strategies KM does not work properly in the business. Construction industry follows some special forms of knowledge management practice. In this case, the professional teams develop national and international expertise in certain areas of work. In construction companies, the term 'knowledge management' is new and it creates debate also, whether it is a kind of passing management fad or it can be a permanent asset to the organization [25]. There are very few companies under the construction industry, which appoint knowledge management officers or experts but others have limited access toward usage of intranets. Basically, knowledge management activities have been promoted for improving performance of construction industry. A knowledge-based strategy allows a framework by which companies can operate and allocate an appropriate agenda. This will also help to establish time frames during a project life cycle.

3.1.9.2 Development of a Strategy

A strategy is always focused on what major plans are to be undertaken for allocating resources. A major kind of undertaking in an industry is to implement knowledge management in a very structured manner. KM and its developing responsive strategies help to assign responsibility in relation to analyzing new issues for people or groups of people. A strategy based on knowledge management within the organization always gives emphasis on setting up clear goals and how these goals can be achieved within the specified time frame. In construction companies, there are number of considerations like (i) which part of construction process generates maximum benefits, (ii) by KM strategy which section of the company can be enabled to get highest advantages, (iii) in what way a problem can be identified, (iv) in what way the systems can be evaluated, (v) how to map organizational business process effectiveness, (vi) determine tremendous benefits for one small area of such a system, (vii) identifying clear methodology for clear life cycle from the phase of capturing data to knowledge retirement, (viii) determine a knowledge management strategy for dealing with obstacles like data validation or limited time frame, etc. [26, 27].

3.1.9.3 Impact on Structure and Working Practices

The construction industry does not follow a strong record of providing value to its employees and their collective or individual contributions. This makes it more difficult to share knowledge. Tacit knowledge is denoted as personal property rather than organizational property. The knowledge-sharing procedure becomes tough due to hierarchical structure of organization and existence of multidisciplinary teams. It is viewed that radical changes in workplace under this industry are not so much desirable. Any task that involves extraordinary efforts will not be widely accepted. Knowledge management practice helps to make it easier and become an integral part for individuals and groups if they want to achieve success.

3.1.9.4 Cultural and Other Barriers

The typical nature of construction companies does not encourage the culture of knowledge sharing. The staffs are very slow in the process of sharing knowledge. To make the organizational performance strong, the strategy of knowledge management and its benefit must be implemented into the culture of organization [28]. Few barriers are there toward successful implementation of knowledge management such as lack of time for sharing knowledge, difficulty in solving large problems, knowledge conversion, multidisciplinary teams, lack of learning, and many more.

3.1.10 Knowledge Management in Education

In research, knowledge management, as a management tool, has generated higher credibility over the past decade. The researchers and academicians debated on characteristics of KM that whether it is a fad or not. In the academic environment, knowledge management is a unique and emerging field. There are many national and international seminars, conferences held on knowledge management in education sector. So many international universities are actively participating in knowledge management-related activities and doing research on it. This is now a very important factor in education sector to discover intellectual power for sharing experiences. It has a great significance and potential in the education sector. Past events help to create new knowledge. Knowledge is such a thing that builds on knowledge base. Human effort is the best source of generating knowledge. This can be developed by conduction of good education activities, research activities, etc. and generating new concepts in the area of interest [29, 30]. All organizations based on knowledge such as research and development centers, industries, higher education from college to universities, all entities are in search of innovative concepts in the area of subject interest. In this case, they are contributing knowledge through different means. They are regarded as "knowledge houses", where knowledge is followed from teachers to students and has a high potentiality in creating knowledge. Figure 3.2 shows the model of knowledge management in education sector. There are important factors

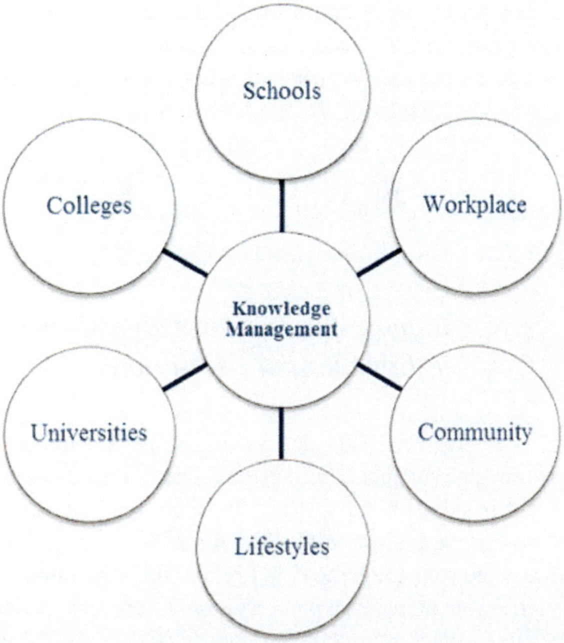

Fig. 3.2 Model of knowledge management in education

included in this model of KM in education such as schools, workplace, community, lifestyles, universities, and colleges.

3.1.11 Knowledge Management in Higher Education

This section establishes some of the important concepts about knowledge management, its processes, structures, systems, and roles in higher education. It sets a vision for the future. At university level, knowledge management activities are very much significant. It has a well-defined use at the foundation for future development. In a knowledge-based society, universities and their staff should recognize and respond toward their changing role. Every university must associate their work process with generation of their knowledge assets. It is important to recognize the value of their intellectual capital to maintain their role continuously in society. These should include all staff and students and not a simple additional burden or agenda set by management. In order to accomplish this process of ownership, the full set of knowledge management will be an important evolutionary process.

The objective of knowledge management in higher education is to create and maintain knowledge repositories, improve accessibility to knowledge, enhancement of knowledge environment, and provide valuation of knowledge. In higher education, knowledge starts with data, raw facts, and numbers as well [30]. An example can be given in this juncture, that is, market value of an institution or a university's endowment, etc. Here information is captured with the help of documents or in the database. With the usage of modern information technology systems, large amount of data can be easily retrieved. When information is combined with knowledge it also generates knowledge. In simple words, knowledge is actionable information [31]. It allows us to make predictions and predictive decisions. Knowledge management has been used and utilized for a long time in higher education. Knowledge is treated as the most valuable asset in the organization.

3.1.12 Importance of Knowledge Management in Educational Institutions

It is established that success of any organization depends on implementation of knowledge management system. There are few key factors of knowledge management in educational institutions such as higher education internalization, lifelong learning, paradigm shift of learning, new unique technologies, and globalization. Knowledge management is able to manage huge amounts of data systematically therefore it becomes a powerful tool to enhance productivity. This enables to reduce cost of collection of a huge quantity of data. To record tacit knowledge is very difficult created by the institutional staff. It has been noticed that the attrition of a staff from an institute leads to loss of knowledge with him. If knowledge management continues to be practiced in an institution then only it can be captured and recorded properly, as well as can be preserved for future use. The inspecting officers while visiting institutes for assessment of the institution's educational development and gradation, reviews all tacit and explicit knowledge of past years with the help of KM tools [32, 33]. Knowledge plays a key role in any type of creation of strategy and decision making. In order to compete and sustain with other organizations in the world, all educational institutes must implement effective system of knowledge management. Knowledge management tools help educational institutions to improve their capability of collecting and sharing data and information. This has an application to problem-solving domain and supports research and development-oriented work. Knowledge management in the educational institutes is comprised of information and shares it from the managerial level to the student level. This provides improvement of professional knowledge of employees and creates quality among lecturers and students. All over the world, government has released many funds for such activities. Knowledge management tools give most effective way to transfer knowledge with attractive methods, models, ideas, and practices. KM creates networking for better interaction between different levels of people in the institutions. It focuses on innovation and development segments as well. In developing countries, faculties can share resources that they invest in mutually. The exchange of information and knowledge can be shared through newsletters, conferences, meetings, seminars, and symposiums. In the knowledge base, students' knowledge, skills, talents must be preserved [34]. This aids in creating new knowledge and KM gives platform to newly enrolled students. In educational institutes, researchers, faculty members, subject experts, students, all contribute knowledge regularly by generating new concepts. Internalization of higher education should be possible through implementation of proper knowledge management practice. Therefore, it can be said that KM helps to capture tacit and explicit knowledge of an educational institution and records it for future use. By the usage of advanced technology, knowledge management effectively transforms new levels in the organization with efficiency and high scope of operation. Data and information are used to improve productivity for users. In the institutions, KM is continuously innovating tacit knowledge. It has a usage in the area of problem-solving and decision-making. An effective knowledge management system in education always tries to achieve improvement.

3.1.13 Knowledge Management in SMEs

Many years ago, it was established that knowledge management practice is very much popular in large-size organizations rather than SMEs. Later the concept of Knowledge Management developed and was applied in small and medium-sized enterprises. The various aspects of knowledge management got attention in SMEs since 2001. This helps in identifying the gap of implementation of KM in SMEs as well [35]. KM is focused on knowledge creation and knowledge transfer within the organization.

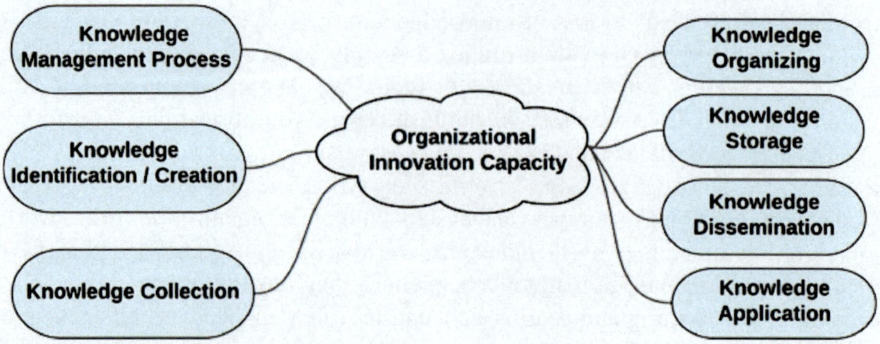

Fig. 3.3 A roadmap of knowledge management on innovation in SMEs

To enhance sales and productivity, organizational culture and learning environment must give attention to knowledge storage and sharing. KM provides the benefits of cost reduction and increase in innovation, creativity, and quality of service. In small and medium-sized enterprises, knowledge management system helps to define the process of business and structure. One of the most important strategic factors of any corporate operations is knowledge. It has the capabilities to achieve a competitive advantage in the market. When KM implementation comes to small and medium-sized enterprises, it gives these firms a power to compete with the large size firms. KM helps SMEs to overcome the limitations due to lack of resources. By doing optimum utilization of the knowledge stock, SMEs can accomplish their short and long-term goals. In many countries, KM practice has become very much significant with other large-scale businesses. Few literatures have established that there is an absence of systematic knowledge management in SMEs due to non-modification in the KM process. Figure 3.3 has shown the roadmap of knowledge management on innovation techniques in small and medium-sized enterprises. Organizational innovation capacity on knowledge management is based on few parameters under SMEs, such as (i) knowledge management process, (ii) knowledge identification and creation, (iii) knowledge collection, (iv) knowledge organization (v) knowledge storage, (vi) knowledge dissemination, and (vii) knowledge application. With the help of these knowledge management stages, SMEs are able to store and transfer valuable organizational knowledge in a creative and innovative way.

3.1.14 Factors Influence KM in SMEs and Its Significance

There are so many factors that influence knowledge management in small and medium-sized enterprises. The factors are, namely, (i) contingency factors, (ii) environmental factors and (iii) historical factors. Contingency factors explain the process of KM when there is a crisis in business operations and KMS application helps to overcome the situation, such as making decisions at the right time and at the right place on critical issues. Environmental factors influence the internal and external environment of businesses. Historical factors are influencing the implementation of knowledge management and its impacts [36, 37]. The success of knowledge management practices in SMEs depends on these factors. Apart from these, some sub-factors are also included in the process such as industrial characteristics and industrial organization, social factors, firm-specific factors, human and cultural factors, technical factors, managerial factors, and many more. In these sub factors, technical and managerial sub-factors are very important. Technical factors denote degree of information technology applications, information systems, and infrastructure. On the other hand, managerial factors define strategy management style on knowledge management, management leadership, infrastructure of an organization, teamwork, and reward system.

It is very essential to know that not only the large-scale organizations are following the practice of KM strategy, small and medium-sized enterprises also follow the strategic implementation of KM. Country's economy is based on productivity and profitability of business organizations. Every business has its objective to make money and profit leading toward business development and growth in future. For achieving this success, business enterprises need to set up a well-defined knowledge management system. SMEs are playing a catalytic role in economic development. In addition, SMEs have the capability to boost today's global economies. Small and medium-sized enterprises represent the promotion of entrepreneurial skills. SMEs are becoming the main source of manufacturing, export, technical innovation, services, employment, etc. According to statistical reports, it is said that 95% of firms come under SMEs. It helps to generate 60 to 70% of the country's employment and 55% GDP as well. In developing countries, the case is little different. Statistics reveal that more than 90% of all the firms, outside agricultural sector, come under small and medium-sized enterprises and micro-enterprises. They generate a significant portion of GDP of those countries. In Bangladesh, 99% of firms consist of less than 100 employees and they generate 58% of employment. In Morocco, 93% of industrial firms come under SMEs and generate 38% of production and 46% employment. It is not true that all micro-enterprises and SMEs come under the formal sector; some are lying under unofficial labor market, which can vary in size, estimated to be about 4 to 6% in developed countries and more than 50% in developing countries. Countries like Australia, New Zealand, etc. have about 90% of businesses under small-scale businesses [38–40].

3.1.15 Knowledge Management in Small and Medium-Sized Enterprises

There are many smaller firms that are facing resource constraints, which in turn creates challenges in their decision-making process. SMEs have flat organizational structure and free-floating management style. This encourages the promotion of entrepreneurship and innovation (informal and non-beaurocratic firms). Owner's personal supervision helps to control the same. There is a lack of formal policies in SMEs. Many smaller firms have owner-managers who hold the central position. In this case, decision-making process and business planning processes all are centralized (to only one person), but this is not an appropriate business practice. Workforce must know about business operations. Therefore, people of the firms should know the benefits of knowledge management and with the help of these, they will be able to enhance, share and transfer knowledge among SMEs. Every business operation requires close attention for its smooth running. If strategy is insufficient and knowledge is poorly managed, businesses face different emerging challenges. Financial resources and expertise are also needed in this case. This generates physically stored knowledge for owners and some key employees [41]. Therefore, it is clear that SMEs are facing some unique KM challenges which are different than large size corporations. Knowledge management processes help SMEs to reduce risks. This provides the benefit of better decision-making process. At operational level, SMEs tend to apply practice of knowledge management which is very much essential. Management should focus on implementation of tacit knowledge and communication channel in this field. In case of SMEs, KM appears as less advanced in terms of construction of knowledge as it has more mechanical approach. For enhancement of tacit and explicit knowledge and making it transferable, SMEs should provide an extreme emphasis on this area with KM strategy. Organizational learning is very crucial in every phase of business operations. Workforce needs to learn new techniques or methodology for doing work faster and smarter. KM strategy of SMEs needs modification and upgrading like large-scale organizations.

3.1.16 Benefits of Knowledge Management in SMEs

There are a lot of significances of KM existing in small and medium-sized enterprises. To find out the way of exploiting knowledge in business processes, SMEs have pivoted on KM strategies. It will make communication between employees easier. The benefits of knowledge management in the field of SMEs are like (i) it influences risk-minimization in organizations; basically, KM captures and locates organizational valuable knowledge that reduces risk factor, (ii) it influences efficiency seekers in organizations—which leads to maximum use of existing knowledge in the corporation by the help of knowledge transfer and sharing practices, (iii) it influences innovators in organizations—most of the time innovators focus on new and unique knowledge and processes that enable creativity and innovation.

The successful agenda of a business always depends on Knowledge Management contribution, for improving competitive advantage of a business in several ways. This includes (i) allowing SMEs to develop a better understanding of customer database and need, preferences of clients, (ii) facilitating long term partnership-style relationships with customers and clients, (iii) contributing to an organization to set up and sustain their status in appropriate leadership style, (iv) improves the speed and quality of business processes, (v) assists SMEs to utilize lessons which are learned from previous jobs, tasks, duties, and projects. This will lead to improve their future performance [42].

The business strategy for SMEs is little bit different from large corporations. The application of knowledge management paradigm in SMEs is very essential for achieving their goals. It depends on nature of projects, plans, procedures, and objectives of the firms [43]. It is seen that, as SMEs have less resources therefore their goals should not be long-termed. On the other hand, as large size corporations have plentiful resources; their business strategies by using KM are extremely focused on long-term and futuristic goals. Most of the time when SMEs have the intention to make strategy based on knowledge management, they try to capitalize less expensive and higher interactive tools and systems. They have pivoted on web-based KM software for documentation, gathering information on customers, suppliers, and employees. It is well established that the knowledge management goals and short-term/long-term strategies of SMEs are very much interconnected.

3.1.17 Knowledge Management in E-business

It is very essential for businesses to set up a strategy which generates high competence level in the market. In the modern era of knowledge-based economy, firms are creating knowledge and transferring this knowledge for value addition. In e-business, knowledge economy presents general motivation that captures knowledge for making innovation in business. Continuous updating in technology is a cause of reduction of product life cycle and need for innovation in e-business. Knowledge management has given birth to sustainable competitive advantage in e-business areas. It has been revealed by research that knowledge management plays a significant role in improving the level of performance of new product development. It is indeed a true statement that knowledge can be created, transferred, and utilized in the organization for generating sustainable competitive advantage in the market. Innovation and knowledge both are interconnected to each other. Innovation is the application toward creating new knowledge. Furthermore, innovation may arise at the intersection between flow of people and flow of knowledge into the organization. Knowledge management helps to share knowledge across geographical and cultural boundaries [44, 45]. E-business is totally based on information and knowledge value chains. The internal employees and external users have given their increasing involvement in e-business process. Knowledge management in e-business is boosting the productivity and creativity. It also facilitates innovation in corporate settings. The under mentioned Figure (Fig. 3.4) has shown KM process reference model in e-commerce. There are few steps in this reference model for business decision-making goals such as (i) problem identification, (ii) data understanding, (iii) method selection, (iv) data reprocessing, (v) pattern discovery, (vi) knowledge evaluation, (vii) knowledge codification, (viii) knowledge repository, and finally (ix) knowledge deployment.

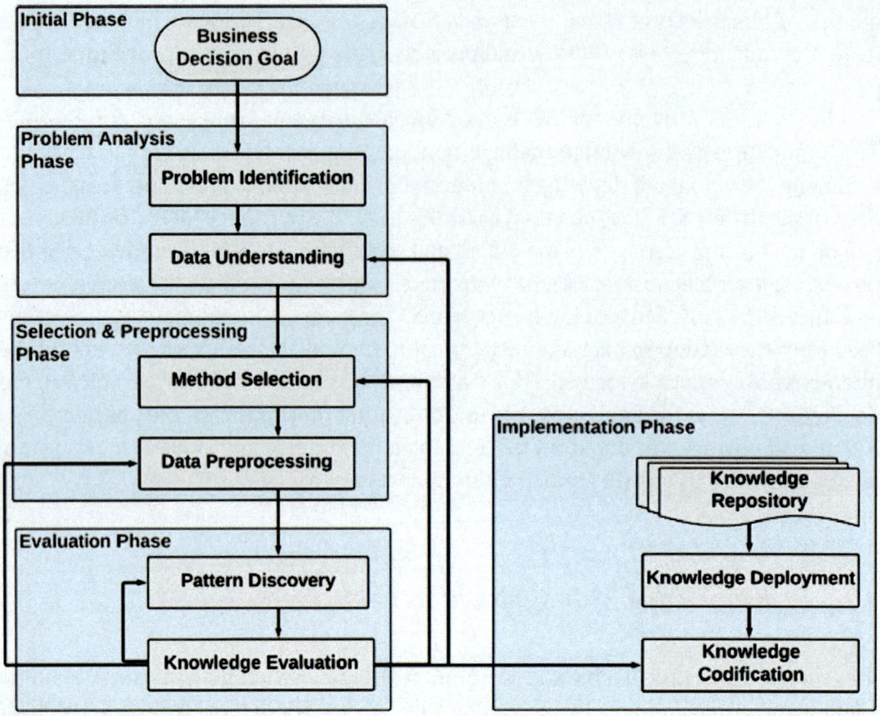

Fig. 3.4 KM process reference model for e-commerce

The knowledge evaluation step influences the above steps in the model name, data identification, method selection, data reprocessing, and pattern discovery [46, 47].

3.1.18 *Significance of Knowledge Management—E-commerce*

At present, e-commerce businesses are highly dependent on qualitative knowledge management. For better customer acquisition, retention, and growth of business knowledge management has played a vital role in e-commerce industry. There are important three reasons to embrace knowledge management for e-commerce industry.

3.1.18.1 Drive Revenue and Profit Growth

E-commerce businesses can develop by using data and information with high-quality knowledge management tools. New market products, launching products in the market, and application of those products to larger prospects can be identified based on these data. The data is able to identify characteristics of products and features which can boost sales. In this way, a positive influence can be created on marketing efforts, i.e., which images can be used in banner ads or e-mail campaigns. If data is available, accurate analysis can easily be made along with decision-making. Trend analysis should be done on advertising and merchandizing in this case [48]. In e-commerce industry, information is priceless. It helps to keep them stay ahead in the market and delivers relevant marketing services to customers.

3.1.18.2 More Accurate Consumer Targeting

For e-commerce businesses, the top priority is customer acquisition and retention. Knowledge management practices help to align advertising efforts to keep existing customers for long time and attract new ones. With the help of different tools of knowledge management segmentation, targeting and position of products and services become easier. For retaining customers, e-commerce businesses also provide discounts on products to repeat customers. Customer base can be generated through KM tools. Customer profile can pull through effective usage of KM also. When it comes to using a new product then larger discount must be offered to customers. Moderate discounts are offered to those customers who have moderate propensity of purchase.

3.1.18.3 Managing Variable Customer Service Expenses

It is most important to maintain high level of customer service for e-business. Knowledge management enables to hire and train customers in e-commerce. This includes integrated call center software. The tools of knowledge management generate higher level of interaction with customers. This is also one important reason for implementing knowledge management in e-commerce [49]. This helps to enhance efficiency of business. With these benefits, a business can expand profits, target customers, and control customer service expenses.

3.1.19 Benefits of KM Tools into E-business Information Management

Knowledge management tools are unique technologies and resources that help to transfer, generate, and codify knowledge easily. All types of knowledge management tools are not computer-based applications. Knowledge can be transferred via phone calls also. There are few advantages of KM tools in e-business information management. Proper organization and evaluation of requirements of customers and suppliers and their inter-relationship is one of them. In this case, questionnaire is a tool for collecting data and information. This is one of the common methods of gathering data and information. Only gathering information is not a new thing for organizations. The classification and evaluation processes are also important that are based on collected information from customers, partners, suppliers, stakeholders, etc. This information is related to satisfaction, suggestion, recommendation, and requirement. Knowledge-based system or KBS is a special KM tool that gives emphasizes on collection of information and its evaluation. The prime aspect of this process is to make correct decisions for organizations in terms of customers' demands or suppliers' preferences. This identification is making "new strategy" for organizational change. One example can be given in this juncture that after classification of knowledge pertaining to customer feedback; it is observed by the organizations that customers are not getting up-to-date information about new products or services, their prices, and offers [50–52]. It gives a food for thought to the organization about new "marketing tools" for improving promotion, namely: e-mail marketing and e-brochures marketing.

3.1.20 The Contribution of KM in E-business Strategy Stages

3.1.20.1 Initiate Stage

First stage is the initiate stage. The goals of this stage are (i) outline scope of the project, (ii) identification of project stakeholders, (iii) determination of scheduling projects. Project scope and scheduling deal with data and information are based on prediction and general studies. For these two tasks, knowledge management is very much significant. This is because both acts depend on gathered data or "declarative knowledge" from customers and suppliers. Behavioral knowledge is also included like organizational documents, competition in market, prediction for deliverables, etc. All tasks can be easily performed by few knowledge managements tools in e-business. They are knowledge mining, organizational knowledge base, knowledge determination, etc. There is also a significant role of knowledge management in the area of identification of project stakeholders' tasks. KM helps to manage relationship between stakeholders. Transfer of knowledge assists stakeholders to create the value proposition. Knowledge repository has enabled the firm for identification of affected customers, suppliers, employees, departments, public as well as private sectors in the new system.

3.1.20.2 Diagnose Stage

The main purpose of this stage is to make SWOT analysis of current business strategy. SWOT analysis means strength, weakness, opportunities, and threat analysis of a business. This is possible by analyzing the position of organization among competitors and revision of current strategy for understanding current relationships between organization and its suppliers or customers. To know about the current position of the organization, there are some analytical tools that help to analyze and assess the position of firms. For example, small-size firms may use industry analysis and medium-sized firms may use supply chain analysis. To know the status of customers and suppliers' relationships there is a tool named Customer/Supplier Life Cycle [53]. It makes evaluation of relationships easier. All tasks can be done in this stage through organizational knowledge base and knowledge repository as it includes current state of organization which means there is no need of collection of data/information and no prediction for future behavior.

3.1.20.3 Breakout Stage

The third stage is breakout stage. The objective of this stage is to set up and deliver new strategies from overall business approaches. It matches the organizational goals. E-business strategy is based on other organization's plans. This is a recommendation to understand the association between the proposed e-business strategy and other adopted policies. The new business practice must be based on diagnosis stage or assessment of organization policy and SWOT analysis, position analysis is also involved in this stage. Proposed strategy and adopted scheme both are intellectual assets of the organization [54]. The project managers' duty in this scenario is to make derivation process of new strategies including allocation of staffs, distribution of tasks, and identification of IT requirements, satisfying SWOT deficiencies, addition or omission of new features for optimization of overall business strategy. Furthermore, the responsibility of project managers has been added and it deals with human dynamics of the corporation. This is very much concerned with the required technology. It is important to mention that building e-business strategy is the process of conversion of organizational intellectual assets into a new road map of organizational requirements. The re-arrangement of corporate business approach is based on information technology infrastructure. The prime responsibility of CKO or chief knowledge officer is converting knowledge into valuable profit or revenue through management and control of organizational intellectual assets. In this case, CKO plays the role of e-business strategy project manager.

3.1.20.4 Transition Stage

In the fourth stage (transition stage), the firms are implementing proposed roadmaps of the new strategy. Here transition denotes the firm is moving from current state to the proposed state. This transition movement is supported by capabilities of the firm and new resources. In this stage, there is a need for gap analysis in order to avoid risks, changes, and conflicts between current and new proposed strategies. The striking role of knowledge management in this stage is very much crucial and valuable. For implementation of new strategy knowledge management is needed to evaluate the abilities of organizations (e-business). Knowledge is special intangible resource of organizations. It comprises of culture of organization, business processes, policy, and human resource experiences. It will provide assessment to the organization by assessment tools like risk analysis or change analysis. It gives the information whether organization has the ability to cope up with changes of new strategies or it has the ability for alteration. To manage these changes effective critical evaluation is made for all areas of changes [55–58].

3.2 Conclusions

It is already established that one of the most important and complex phenomena in business is knowledge. It has a positive impact on the life of individuals, organizations, and the whole existence of society. In various sectors, knowledge management practices can act as a source of change and development. Proper implementation of knowledge management in different sectors leads to better life for working people and higher achievements for the business houses. Knowledge is such a thing that helps organizations to expand new domains, enhance activities, and improve organizational culture and development. On the other hand, these different sectors, with a great practice of knowledge management, have generated a steady progression in the society on economic and technological aspects. Moreover, it can be said, envision of knowledge processes are fundamental drivers of life of individuals on all levels.

References

1. Gupta, B., Iyer, L. S., & Aronson, J. E. (2000). Knowledge management: practices and challenges. *Industrial Management & Data Systems.*
2. Inkinen, H. (2016). Review of empirical research on knowledge management practices and firm performance. *Journal of Knowledge Management.*
3. van Beveren, J. (2003). Does health care for knowledge management? *Journal of Knowledge Management*.
4. El Morr, C., & Subercaze, J. (2010). Knowledge management in healthcare. In *Handbook of research on developments in e-health and telemedicine: Technological and social perspectives* (pp. 490–510). IGI Global.
5. Kothari, A., Hovanec, N., Hastie, R., & Sibbald, S. (2011). Lessons from the business sector for successful knowledge management in health care: A systematic review. *BMC health services research, 11*(1), 1–11.
6. Orzano, A. J., McInerney, C. R., Scharf, D., Tallia, A. F., & Crabtree, B. F. (2008). A knowledge management model: Implications for enhancing quality in health care. *Journal of the American Society for Information Science and Technology, 59*(3), 489–505.
7. Gowen, C. R., III., Henagan, S. C., & McFadden, K. L. (2009). Knowledge management as a mediator for the efficacy of transformational leadership and quality management initiatives in US health care. *Health care management review, 34*(2), 129–140.

8. Berler, A., Pavlopoulos, S., & Koutsouris, D. (2005). Using key performance indicators as knowledge-management tools at a regional health-care authority level. *IEEE transactions on information technology in biomedicine, 9*(2), 184–192.
9. Lobach, D., Sanders, G. D., Bright, T. J., Wong, A., Dhurjati, R., Bristow, E., Bastian, L., Coeytaux, R., Samsa, G., Hasselblad, V., Williams, J. W., Wing, L., Musty, M, Kendrick, A. S. (2012). Enabling health care decisionmaking through clinical decision support and knowledge management. *Evidence Report/Technology Assessment*, (203), 1–784
10. Bordoloi, P., & Islam, N. (2012). Knowledge management practices and healthcare delivery: A contingency framework. *The Electronic Journal of Knowledge Management, 10*(2), 110–120.
11. Sibbald, S. L., Wathen, C. N., & Kothari, A. (2016). An empirically based model for knowledge management in health care organizations. *Health care management review, 41*(1), 64–74.
12. Bate, S. P., & Robert, G. (2002). Knowledge management and communities of practice in the private sector: Lessons for modernizing the national health service in England and Wales. *Public Administration, 80*(4), 643–663.
13. Corrao, S., Arcoraci, V., Arnone, S., Calvo, L., Scaglione, R., Di Bernardo, C., Lagalla, R., Caputi, A.P., Licata, G. (2009). Evidence-based knowledge management: an approach to effectively promote good health-care decision-making in the Information Era. *Internal and Emergency Medicine, 4*(2), 99–106.
14. Hendriks, P. H., Ligthart, P. E., & Schouteten, R. L. (2016). Knowledge management, health information technology and nurses' work engagement. *Health care management review, 41*(3), 256–266.
15. Kamara, J. M., Augenbroe, G., Anumba, C. J., & Carrillo, P. M. (2002). Knowledge management in the architecture, engineering and construction industry. *Construction Innovation.*
16. Rezgui, Y., Lima, C., Wetherill, L., & Zarli, A. (2002). Knowledge Management for the Construction Industry: The E-Cognos Project'. *Electronic Journal of Information Technology in Construction*, 183–196.
17. Rezgui, Y. (2001). Review of information and the state of the art of knowledge management practices in the construction industry. *The Knowledge Engineering Review, 16*(3), 241–254.
18. Rezgui, Y., Hopfe, C. J., & Vorakulpipat, C. (2010). Generations of knowledge management in the architecture, engineering and construction industry: An evolutionary perspective. *Advanced Engineering Informatics, 24*(2), 219–228.
19. Kazi, A. S. (Ed.). (2005). *Knowledge management in the construction industry: A socio-technical perspective.* Igi Global.
20. Zhang, X., Mao, X., & AbouRizk, S. M. (2009). Developing a knowledge management system for improved value engineering practices in the construction industry. *Automation in construction, 18*(6), 777–789.
21. Fong, P. S., & Choi, S. K. (2009). The processes of knowledge management in professional services firms in the construction industry: a critical assessment of both theory and practice. *Journal of Knowledge Management.*
22. Yu, D., & Yang, J. (2018). Knowledge management research in the construction industry: A review. *Journal of the Knowledge Economy, 9*(3), 782–803.
23. Maqsood, T., & Finegan, A. D. (2009). A knowledge management approach to innovation and learning in the construction industry. *International Journal of Managing Projects in Business.*
24. Navarro, J. G. C., Dewhurst, F. W., & Eldridge, S. (2010). Linking chief knowledge officers with customer capital through knowledge management practices in the Spanish construction industry. *The International Journal of Human Resource Management, 21*(3), 389–404.
25. Svetel, I., & Pejanović, M. (2010). The role of the semantic web for knowledge management in the construction industry. *Informatica, 34*(3).
26. Boamah, F. A., Zhang, J., Wen, D., Sherani, M., Hayat, A., & Horbanenko, O. (2021). Enablers of knowledge management: practical research-based in the construction industry. *International Journal of Innovation Science.*
27. Christiansson, P. (2003). Next generation knowledge management systems for the construction industry. *CIB REPORT, 284*, 80.

28. Fargnoli, M., De Minicis, M., & Di Gravio, G. (2011). Knowledge Management integration in Occupational Health and Safety systems in the construction industry. *International Journal of Product Development, 14*(1–4), 165–185.
29. Khatun, A., George, B., & Dar, S. N. (2021). Knowledge Management Practices in the Higher Education Sector with Special Reference to Business Schools. *Education and Self Development, 16*(2), 47–59.
30. Edge, K. (2005). Powerful public sector knowledge management: a school district example. *Journal of Knowledge Management.*
31. Massaro, M., Dumay, J., & Garlatti, A. (2015). Public sector knowledge management: a structured literature review. *Journal of Knowledge Management.*
32. Sallis, E., & Jones, G. (2013). *Knowledge management in education: Enhancing learning & education.* Routledge.
33. Kidwell, J. J., Vander Linde, K., & Johnson, S. L. (2000). Applying corporate knowledge management practices in higher education. *Educause quarterly, 23*(4), 28–33.
34. McGinn, N. (2001). Knowledge management in the corporate sector: Implications for education. *Development Knowledge, National Research and International Cooperation.* CAS/DSE/Norrag, Bonn and Edinburgh.
35. Durst, S., & Edvardsson, I. R. (2012). Knowledge management in SMEs: a literature review. *Journal of Knowledge Management.*
36. Desouza, K. C., & Awazu, Y. (2006). Knowledge management at SMEs: Five peculiarities. *Journal of Knowledge Management.*
37. Hutchinson, V., & Quintas, P. (2008). Do SMEs do knowledge management? Or simply manage what they know? *International Small Business Journal, 26*(2), 131–154.
38. Cerchione, R., & Esposito, E. (2017). Using knowledge management systems: A taxonomy of SME strategies. *International Journal of Information Management, 37*(1), 1551–1562.
39. Salojärvi, S., Furu, P., & Sveiby, K. E. (2005). Knowledge management and growth in Finnish SMEs. *Journal of Knowledge Management.*
40. Cerchione, R., Esposito, E., & Spadaro, M. R. (2015). The spread of knowledge management in SMEs: A scenario in evolution. *Sustainability, 7*(8), 10210–10232.
41. Handzic, M. (2006). Knowledge management in SMEs. *CACCI Journal, 1*, 1–11.
42. Nunes, M. B., Annansingh, F., Eaglestone, B., & Wakefield, R. (2006). Knowledge management issues in knowledge-intensive SMEs. *Journal of Documentation.*
43. Lee, M. R., & Lan, Y. C. (2011). Toward a unified knowledge management model for SMEs. *Expert systems with applications, 38*(1), 729–735.
44. Malhotra, Y. (2000). Knowledge management for e-business performance: Advancing information strategy to "internet time." *Information Strategy: The Executive's Journal, 16*(4), 5–16.
45. Rowley, J. E. (2002). Reflections on customer knowledge management in e-business. *Qualitative Market Research: An International Journal.*
46. Maguire, S., Koh, S. C. L., & Magrys, A. (2007). The adoption of e-business and knowledge management in SMEs. *Benchmarking: An International Journal.*
47. Lin, H. F., & Lee, G. G. (2005). Impact of organizational learning and knowledge management factors on e-business adoption. *Management Decision.*
48. Koh, S. C. L., & Maguire, S. (2004). Identifying the adoption of e-business and knowledge management within SMEs. *Journal of Small Business and Enterprise Development.*
49. Chong, A. Y. L., Ooi, K. B., Bao, H., & Lin, B. (2014). Can e-business adoption be influenced by knowledge management? An empirical analysis of Malaysian SMEs. *Journal of Knowledge Management.*
50. Allard, S., & Holsapple, C. W. (2002). Knowledge management as a key for e-business competitiveness: From the knowledge chain to KM Audits. *Journal of Computer Information Systems, 42*(5), 19–25.
51. Egbu, C. O. (2008). Knowledge management for improved construction e-business performance. *J. Anumba, C., et al*, 222–234.

52. Scott, J. E. (2003). The role of trust in e-business knowledge management. *International Journal of Electronic Business, 1*(2), 187–210.
53. Bernuy, A. E. (2011). Knowledge Management and e-Business. In *KMIS* (pp. 337–340).
54. Roy, T. K., & Stavropoulos, C. (2007). *Customer knowledge management (CKM) in the e-business environment: cases from Swedish banks.*
55. Arias-Pérez, J., López-Zapata, E., & Echeverri-Bedoya, A. (2020). E-business capabilities and innovation performance: The mediating effect of knowledge management strategies. *Serbian Journal of Management, 15*(2), 221–239.
56. Georgopoulos, N. B. (2005). Knowledge management as an integral part of strategic management in e-business era. *International Journal of Knowledge and Learning, 1*(4), 373–387.
57. Singh, A., Sharma, A., & Dey, N. (2015). Semantics and agents oriented web personalization: state of the art. *International Journal of Service Science, Management, Engineering, and Technology (IJSSMET), 6*(2), 35–49.
58. Bureš, V., Tučník, P., Mikulecký, P., Mls, K., & Blecha, P. (2016). Application of ambient intelligence in educational institutions: Visions and architectures. *International Journal of Ambient Computing and Intelligence (IJACI), 7*(1), 94–120.

Chapter 4
Knowledge Management System

4.1 Introduction

In any business organization, knowledge is a crucial asset. Throughout an organization, it should be created, shared, and used effectively. The knowledge management process has an intention toward organizing relevant experience whenever it is required. Under the field of knowledge management, it has been established that organizing knowledge helps to achieve the desired goals with the best possible outcome. Knowledge management and knowledge management systems both have a deep relation; both concepts provide lot of significance in a business. There is a slight difference between these two terms. The concept of knowledge management has a multidisciplinary approach, and it is a formal technique to store knowledge [1, 2]. A company can get significant advantage by using knowledge management concept on installing a knowledge management system; this benefits the employees and the shareholders a lot. Knowledge management is a technique that enables people to get readily available knowledge and information regarding business functions. On the other hand, there is a high requirement for application of knowledge management systems in the workplace for creating, capturing, storing, and distributing information. A knowledge management system helps people in the organization to utilize knowledge effectively to achieve tasks. A knowledge management system reframes the process and creates a most proactive form of work to make a customer attracted toward the firm's product and service. Through this system, customer queries can be answered in real-time; this system saves time from constantly replying to the same questions by the customers. A knowledge management system emphasizes the domain of a well-organized and up-to-date process of work being done in a business. An effective knowledge management system delivers information to the internal as well as external people who are in need of such information [3]. The system is not only good for internal business; it is suitable for customers also. When customers ask too many questions to the service providers, and they have to give the answer accordingly, it consumes a long time for explaining things to each customer. Therefore, the installation of an appropriate knowledge management system helps make their day

S. Majumder and N. Dey, *AI-empowered Knowledge Management*,
Studies in Big Data 107, https://doi.org/10.1007/978-981-19-0316-8_4

easy to handle customers. KMS focuses on the improvement of customer experience in three different ways, such as (i) reducing wait times, (ii) providing more detailed information to internal and external parties of business, and (iii) making customers enable to help themselves. A KMS is a very much helpful technique to manage knowledge externally and internally. The employees of the organization access internal knowledge; internal information is not accessed by customers, i.e., policies, programs, a requirement of HR materials, etc. If internal knowledge is managed properly, it can significantly impact the external knowledge base by providing strong customer support activities [4]. Figure 4.1 shows the various activities of a knowledge management system in an organization namely: (i) accumulation of information, (ii) gathering of information, (iii) identification of knowledge, (iv) storage of knowledge, (v) propagation of knowledge, (vi) accessibility of knowledge, (vii) supply of knowledge, and finally (vii) systematization of knowledge. An organization can compete with other firms in the market by optimum utilization of knowledge management platforms. There is much more established software that has come under the KMS portal. When the organization is thinking and providing emphasis on the effectiveness of knowledge management in business, they must.

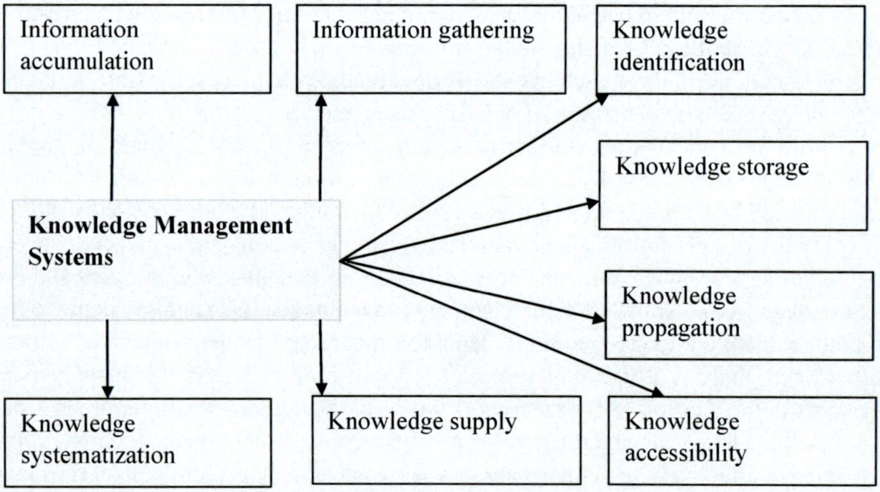

Fig. 4.1 Knowledge management system

4.2 Features of Knowledge Management System

4.2.1 *Ease of Use/Adoption*

The knowledge management system is user-friendly. It minimizes disruption and provides value to the end-users. The various functional teams in the workplace can handle it very easily by importing and formatting content, information, or data.

4.2.2 *Intelligent Integration*

This is a system that can easily capture and store team members' knowledge and expertise. The system includes every important interaction and web search of content. The process of intelligent integration can be done easily with the help of an appropriate knowledge management system [5, 6]. It provides the team members the ability to integrate with one another. The members of the team can use and sync the content of internal and external business climate, with proper storage, verification, and access to knowledge in one place.

4.2.3 *Organization*

Another feature of knowledge management system is the organization of information. Collection of information, grouping, tagging, etc. become easy for users. The system determines who can see what and how an intuitive search can be done.

4.2.4 *Accessibility*

Employees are the critical asset of any organization; their performance will always be counted in the path of accomplishing organizational goals. Employees must have easy access to knowledge for doing their job better. There is no limitation to what kind of devices they are using during the work, but here, the quality of the work is important. KMS can work effectively when accessibility of knowledge gets priority in the workplace.

4.2.5 Customization

The knowledge management system is a device that has a content-focused customization feature that allows people to modify specific suggestions on knowledge management as per their requirements [7].

4.2.6 Collaborative Features

KMS helps to connect team members with one another, and thus they can share their expertise. This system is essential for effective collaboration between functional teams. It enables people by providing the right knowledge to the right people at the right time and at the right place. By this collaborative process, the working people can utilize and share their knowledge or subject matter of expertise on the same page of the organization [8].

4.2.7 Content Verification

There are various powerful knowledge management features under the system as an alert for duplicate content, expert verification, inaccurate content, suggestion for tags; this gives an improvement in knowledge by artificial intelligence integration or AI-based technology.

4.2.8 Smart Suggestions

Employees in an organization are always busy having a conversation with team members, customers, on calls or any chat tools in real-time. The KMS makes it easy to share and transfer knowledge within the tentative time frame. This process suggests specific information before the requirement of data. This nature of KMS encourages them to adapt and fit with the system more.

4.3 Types of Knowledge Management System

The content of knowledge management system includes the key factors of knowledge capital. People should identify the need for a knowledge management system in an organization; they can adapt the new platform to enhance their skills and knowledge. For better understanding of internal and external customers, every firm should focus

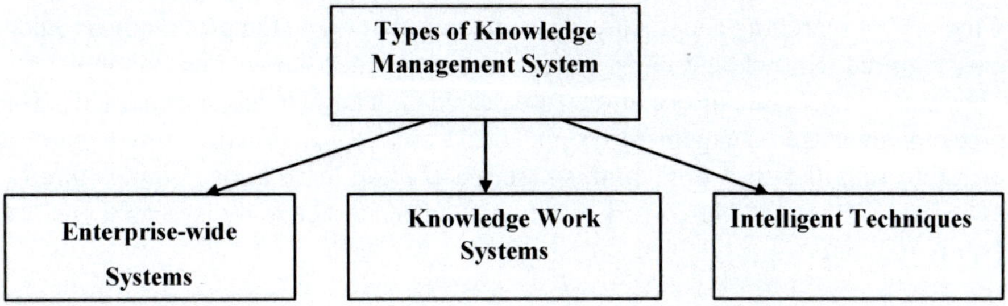

Fig. 4.2 Types of knowledge management system

on using knowledge management systems and various tools. This system increases the potentiality of knowledge resources within a person. On the basis of organizational culture, the process of knowledge management has been implemented. To preserve miscellaneous information, documentation KMS is very much helpful for the internal and external users. In the advanced, technological world, using the internet, extranet, website –content creation and managing, etc. are the key feature of effective KMS. There are three types of knowledge management systems in an organization; which have enhanced the strength of business and level of productivity (Fig. 4.2). The three types are as follows:

4.3.1 Enterprise-Wide Systems

A general-purpose system is known as an enterprise-wide system. Its work is to collect, store, distribute and use digitalized information and related knowledge. This knowledge has a relation with organizations' learning process. Portals, search engines, various tools for collaboration, learning management systems are the different examples of enterprise-wide systems and technologies [9, 10]. This system focuses on locating knowledge workers with specialized and experienced professionals with the help of tools, online directories, and other details. The enterprise-wide knowledge management system is responsible for searching the relevant information and handling it properly. It delivers knowledge to every individual within the enterprise and stores data in a structured or unstructured format.

4.3.2 Knowledge Work Systems

The other type of knowledge management system is the knowledge work system. It is a special type of knowledge management system that has been designed for scientists, engineers, and other knowledgeable individuals. This system includes an online directory for searching a company's specific information by the professionals whenever required. Knowledge work system provides expert knowledge to employees. They can find accurate information from the existing knowledge database [11]. This type of knowledge management system always creates best practices for the purpose of future activities of a firm. In this case, people can learn a lot of things through experience. The knowledge work system includes interactive tools, search engines, user portals, etc.

4.3.3 *Intelligent Techniques*

One of the best types of knowledge management systems is Intelligent Techniques. This system strategically collects, stores, and manages knowledge in an organization. The Intelligent Technique system deals with particular and predictable tasks in relation to knowledge-based software applications. All users and various business processes can work on this system. The raw data can be converted into valuable knowledge management systems through the usage of artificial intelligence. Knowledge capture, transfer, storage, and acquisition of all functions can be made under this type of KMS. Intelligent techniques in knowledge management systems have used searching tools like machine learning, deep learning, and various big data techniques due to the enhancement of connections. Data can be retrieved very quickly by this type of KMS mechanism [12]. Intelligent techniques allow the company to manage knowledge in a better way and optimization other resources.

4.4 Building Effective Knowledge Management System

A few stages under the knowledge management system can make the process highly effective in an organization. The stages are discussed briefly (See Fig. 4.3).

4.4.1 *Identify and Define the Goals of Your Knowledge Management System*

A company always wants to implement a good knowledge management system and such a KMS must support the working class. It is prime and very important to know about the company's goals and objectives. The company must identify the needs of the application of KMS on the basis of customer support, empowerment, user-friendly workflow, collective knowledge strategy, and brand identification [13]. To solve the problems among customer support teams, KMS has provided suggestions, and enhanced customer experiences a lot. It has also emphasized on empowerment of

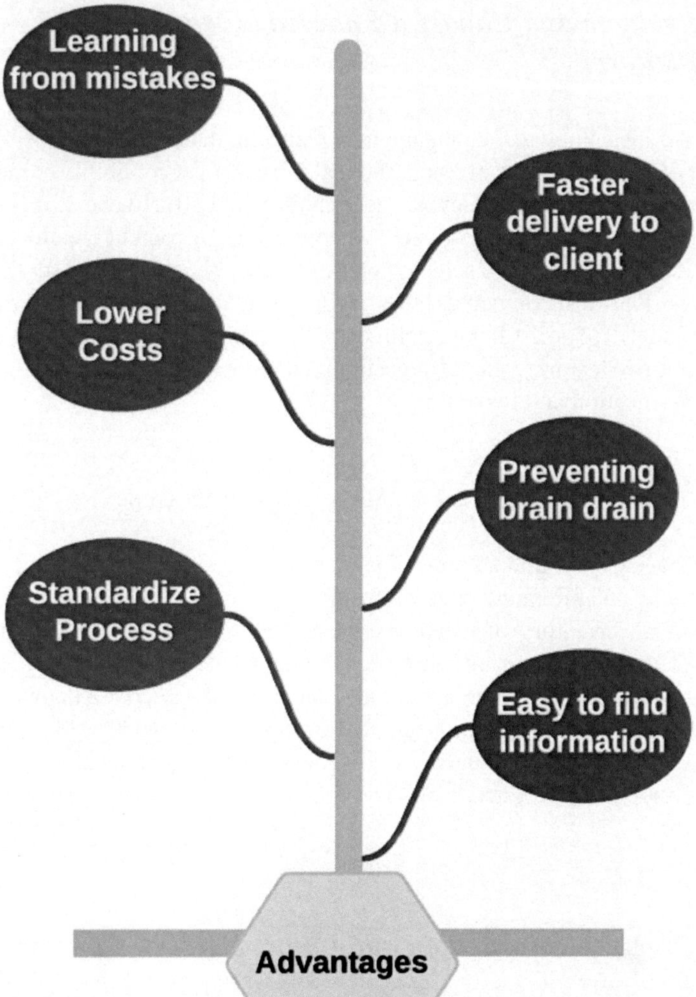

Fig. 4.3 Advantages of knowledge management systems

sales teams by creating shortened sales cycles, keeping up with fast-paced products, and finally converting prospects into satisfied and happy customers. KMS helps to coordinate with the sales team with greater accessibility. It also concentrates on effective knowledge base that must be actionable. It is creating user-friendly workflows for remote employees without knowing the fact where and when they work. The knowledge management system highlights business activities and saves time from answering questions and searching for internal information.

4.4.2 *Evaluate and Choose a Knowledge Management Platform*

Before selecting a knowledge management platform, the evaluation must be there in respect of the user and mode of usage and making sure of accessibility of the platform to any remote employee. To increase accessibility and effectiveness, it is important to choose a platform by teams or users when it comes to content creation, searching internal editing features. The selected platform must be integrated with the existing tools and workflows of companies [14]. It is done to make the platform more adaptable for the employees. The artificial intelligence-driven knowledge management system has collaborative features, smart suggestions, actionable knowledge, which increases AI aided verification and feasibility.

4.4.2.1 Inventory Existing Information and Identify Gaps

The third stage of building a knowledge management system is to identify the most important and suitable knowledge of business and to implement them accordingly. This means an inventory of existing content and recognizing the gaps in knowledge. After that, it focuses on what way the knowledge has been created, shared, updated, and applied in the organization. It is necessary to create knowledge maps with team leaders and key employees by showing what knowledge the team already has and what is required to know for job roles, training and development, competencies development, product design and processes, and strategy formulation of the organization.

4.4.3 *Organize Information and Create Net New Content*

It is well established that in the process of knowledge management system, there must be an appropriate organization that can work on existing content of knowledge and should update the content over time; it functions better for the business firms. Sometimes organization structure is dependent on the needs, demands, and expectations of employees as well as its customers. The usability of the information and content is very much significant for a business organization. KMS helps to make the content more relevant and organized by categorizing topics and adding necessary links within the content. Users can access it most frequently, and content is also logical in this case. The searching options for the next function becomes easier in the area of functional organization, a tool for feedback can create a better organizational structure. In this way, the right kind of people can access information at the right place and at the right time. KMS enables preserving productivity and saving time. The adaptability of team members to the knowledge management system is wider and encouraging. In this way, the continuous update and use of KMS are really incredible.

4.4.4 Implement the Knowledge Management System

The fifth stage of knowledge management system is implementation. For appropriate implementation of KMS, the organization should follow some norms. These are (i) define company's wide objectives and goals and motivations for establishing knowledge management strategy, (ii) regular basis communication on knowledge management updates, (iii) ask feedback from employees, ensure their engagement and improve continuously (iv) fix up various issues quickly so that users can understand input matters, (v) advantage of AI-driven functions like expert verification, flagging inaccurate and duplicate information, understanding and evaluation of users efficiency easily, (vi) rewarding employees for content creation, (vii) a regular basis training conduction on the usage of knowledge management system toward functional teams.

4.4.5 Evaluate and Optimize KMS Performance Post-launch

After launching KMS, evaluation of its effective working and output is required. It is already mentioned that there is a need for regular communication among the users about its usage in the workplace or how much it benefits. If employees are encouraged by an effective knowledge management system, it provides the motivation to continue with this technology. The feedback mechanism helps to ensure a continuous improvement of the system. It gives knowledge of adding further features to a system to make it highly usable to the organization. Knowing the features of the knowledge management system and how it performs in the organization is an important recording phenomenon [15]. The satisfaction level of employees after using knowledge management systems is also recorded. AI-powered technology should be encouraged and used by all levels of employees in different segments of a company. This provides value creation and enhancement to the system [16].

4.4.6 Continue to Improve and Update the Knowledge Management System

Only implementing a knowledge management system is not sufficient to make an effective knowledge-based organization; there is a need to follow up. Continuous improvement and updated knowledge have created an impactful AI-driven system for internal and external customers. In the area of knowledge management system roadmap, continuous improvement and evaluation are necessary over time. This KMS has enhanced a better organizational process in relation to internal and external factors. The need for information among people in the workplace has been satisfied by effective use and modification of knowledge management systems. It has influenced the evolution of knowledge management strategy. The economic development of the organization can take place smoothly and create value addition to the entire system.

4.5 Benefits of Knowledge Management Systems

There are so many essential benefits of a knowledge management system that provides easy dealing and managing organizational information and knowledge capital. The major advantages of the knowledge management system have been discussed below.

4.5.1 Lower Costs

An organization contains appropriate business decisions, business guidelines, and a large workforce. It leads to time-saving operations and provides cost-benefit to the organization when a knowledge management system has already been implemented in the workplace. This becomes possible due to the inclusion of database and knowledge management programs in the bottom line of a business firm. Therefore, the use of knowledge management system has made the organizational work easy and cost-effective [17, 18].

4.5.2 Improved Decision-Making Power

A knowledge management system must focus on improving the decision-making ability of a firm. Sharing knowledge in a period of scarcity is very much essential to accomplish the objectives of the business [19]. This can be done through searching, syndication, backup services, formulation of structure, etc. With the help of knowledge management support, various actions can easily be accessed such as improving product quality, increasing the problem-solving ability, accessing market conditions. Real-life work experiences can fluctuate, thus, decision-making power is increasing continuously for a business house.

4.5.3 *Easy to Find Information*

A knowledge management system makes finding information easier. An individual can search for metadata, templates, and other benefits. It is easy for the people who work in the organization, to access the information in real-time. Knowledge sharing and transfer processes can make the information available quickly. Everyone can perform their job efficiently.

4.5.4 *Tracking All Ideas, Documents, and Other Data*

Tracking ideas, documents, and essential factors are the most important benefit of a knowledge management system in any business house [20]. The working people can use these documents in an effective manner to make the firm productive. There is a lot of knowledge in different business domains, and it needs to be stored and managed properly for future use. KMS has helped a lot for tracking ideas, information, and documentation.

4.5.5 *Preventing Brain Drain*

The knowledge management system has given full insurance to the users. It prevents brain drain. Employees who are working for the organization are able to enhance their skills, knowledge, and competencies by using the different aspects of KMS [21, 22]. They become resourceful and are capable of solving any kind of business issue. By enhancing talent, they can easily store and retain knowledge for the business.

4.5.6 *Learning from Mistakes*

The collective work knowledge has made the work error-free. Learning management system gets new learners more engaged and proactive in relation to their new knowledge, skill, talent, intelligence competencies, and many more. If there is teamwork in the organization, collective knowledge completes the work easily. They might get lessons from various resources [23]. The level of potentiality can increase tremendously. One example that can be given at this juncture is how NASA has learned the reason behind space shuttle explosions and how to prevent such recurrences.

4.5.7 Faster Delivery to Clients

The knowledge management system delivers various benefits to the clients very fast. Among competitors, the speed of execution is one of the most significant differentiators. The organization believes that communication and conveying messages faster will give them a cutting edge over others. It is well established that data sharing, recycling and advancement can take less time to convey any proposition, item, information to a client in the market. KMS provides emphasis on new clients networking, problem-solving ability, add-on benefits to business and capital management, and resourceful collaboration.

In any business house, employees or human assets must stay happy and productive. The external customers also need to stay in happy state of mind after using a company's product. They can show trust and faith in the company and its products as well. In this way, the system can insist them to purchase a company's product multiple times from multiple locations. Therefore, it can be seen from the above discussions that a knowledge management system provides lot of advantages in business proposals as well as toward business productivity [24–26].

4.6 Examples of Knowledge Management System

There are several softwares available on knowledge management systems that are used by different business organizations. Few software-based applications are discussed below (Fig 4.4).

4.6.1 CloudTutoria

The special feature of cloud tutorial is that it supports options on customization. It helps to maintain team members internally as well as externally. The software is a very impressive program for data analysis purposes and is able to report the data functionality. It is full of efficient content management systems. This provides answers for the frequently asked questions by customers and allows the customers to find out their answers quickly. In this way, it helps to save time for business operations. This is basically a knowledge-based platform that can save a firm's manpower hours. This knowledge management system gives the facility for g categorization of the data. By use of it, anyone can efficiently manage general information. This has an impact on users. It is a very easy-to-use content management system [27]. All a to z data models can be managed through cloud tutorials [28].

Fig. 4.4 Examples of knowledge kanagement system

4.6.2 *Tettra*

Tetra is another type of knowledge management system. Some special features exist in this application, such as it creates unbound integration and notification. It acts as a simple editor. Internal page linking function becomes easy through Tettra. It generates an automated table of content [29, 30]. Tettra is a knowledge management system that maintains workflows internally. This offers inspiring answers to questions that are repetitive in nature. This software helps to create essential policies and procedures under a single centralized platform. The process of documentation gets easy. Sometimes it is called a knowledge repository.

4.6.3 *Document 360*

Document 360 is another example of knowledge management system. This has the ability to create, collaborate and publish a self-serving knowledge management system. For managing multiple documents, it supports the SaaS platform. Advanced analysis and reporting tasks can be done with the help of document 360 software. It can maintain multiple versions. The special KMS document 360 has generated a unique platform by an interactive end-user interface. It helps to maintain various third-party integrations. The basic notion of this application is appropriate planning, evaluation of business values, measurement of various benefits, minimize new rollout. The potentiality of success in knowledge management is increasing day by day by using document 360. It comes through an influential artificial intelligence-based searching [31]. This system assists customers by discovering solutions to difficult problems. At any time, by implementing document 360, firms can manage and store files, videos, etc.

4.6.4 Zendesk

Zendesk is one of the best knowledge management systems, among others. Zendesk has been designed for customer engagement; apart from that, it has been used for sales and support purposes. Customers' interaction, tracking of information and prioritization, etc. can be effectively solved by this type of KMS [32, 33]. All users in the organization use such a knowledge management system. This has become an ideal selection for all business firms irrespectively of their sizes. Educators, IT professionals, HR professionals all get benefits by using it. Some of the important features of Zendesk are (i) it is easy to implement, (ii) for better customer support provides effective chat communication, (iii) very much responsive for better understanding, (iv) the feedback mechanism process is smooth) it supports the transaction process, including supporting tickets. This knowledge management system is used for both public and private knowledge bases and serves customers and employees as well [34].

4.6.5 KnowAll

KnowAll is such a knowledge management system that has a wide-ranging solution for enterprises. The best benefit of implementing KnowAll in the workplace is its easy usability. It is basically playing a melody for WordPress. A knowledge base can be formulated in less time, and customization is possible for aligning it into the brand. With the help of KnowAll, any new articles can be built easily with some clicks, and after that, these can be established into various groups and subgroups. The theme of articles can produce manifold designs to cater to a firm's requirements under the terms of interface. This KMS is suitable for supporting customers and their education. The essential features of KnowAll software are, (i) can generate numerous sections, subsections for establishing articles and their execution, (ii) installing themes process is easy and fast to publish articles, (iii) ease in users' feedback on all articles and analytics, (iv) as per the branding, generating numerous design alternatives and modifying them easily[35].

4.6.6 Zoho Wiki

Zoho is another type of knowledge management platform. It acts as a reasonable business solution tool. This software is not so much expensive though the tools are elementary. It is similar to widespread software. Zoho Wiki has been used by various business firms nowadays [36]. This knowledge management system has uncomplicated features. It is like wiki centered knowledge base. Most of the time it is used in small businesses. To organize the content properly it has been categorized through data and information. Content management process has become very easy with this knowledge management platform. Different kinds of content can be managed and created in less time. Zoho Wiki supports other zoho products with several services. This system can filter which groups and individuals may view this different content in the business firm [37].

4.6.7 HubSpot

One of the most well-known customer support software is HubSpot. This is a help desk toward knowledge management database. It gives the benefit of live chat. The user feedback mechanism has made the software effective and established. This provides a unified view and platform of consumer communication. The fabricate articles process has been generated very smoothly by using this software. The frequently asked question or FAQ is converted into a searchable library with so many videos, articles, and documentation. It is a special type of knowledge-based system for huge or large organizations [38]. It has some important attractive features, (i) it is very simple to use, (ii) creation of articles or templates is an easy procedure, (iii) automatically can import existing knowledge base articles, (iv) the knowledge base page can be categorized and adjusted by footer, header, and appropriate design, (v) customers can get feedback in their desired language, (vi) examines the effort of organizational people to identify the need of generating new articles and how requirement can be created for producing new articles (vii) performance of articles and video engagement process are being checked properly by HubSpot.

4.6.8 Knowmax

Knowmax is also a knowledge management software that is enterprise-oriented. It helps to desseminate exact knowledge in the organization. It includes decision-making trees, document management tools, central content management and guides the way of assisting in the production of actionable content. This platform is artificial intelligence-aided search engine, it can facilitate the distribution of contextual knowledge in a correct channel at the right place and at the right time. It is a very simple software to collaborate with agents. Knowmax defines the organizational culture with knowledge transforming and sharing among others. This system acts as a resolution center of knowledge management with an endorsing customer self-service. Basically, it provides cutomer support services for better management of external relations [39, 40]. It gives the view of 360-degree analysis by the omnichannel communication. This system consists of rational decision trees. The customer support services under this software have been based on artificial intelligence.

4.7 Case Studies

4.7.1 Knowledge Management at Infosys

It is very interesting to study knowledge management (KM) practices in the information technology sector. In this study, we have focused on KM at Infosys (a global leader in next-generation digital services and consulting), based on the secondary data available over the internet. There are various types of knowledge dissemination initiatives that can be shared, reused, and decentralized for companies' better performance. The company can manage and arrange a knowledge body based on intranet by conducting experimental learning and earning through previous projects. The giant company like Infosys has maintained a well-established practice of knowledge management. A virtual classroom (intranet-based) is used to access various internally developed (in-house) course contents besides incorporating a forum of discussion. In this forum, the participants can actively participate to post and respond to raised queries related to the courses. The company-wide intranet is full of 5000 nodes and is spread out over all India-based development centers (DCs) and marketing offices in US. The KM practice in Infosys has been continued through regular seminars and sessions. These programs are held within units and as well as throughout the organization. Infosys has moved toward an integrated Knowledge Management strategy since 1999. The leading brand under IT sector, Infosys, focused on people, processes, and technology for enhancement of their KM strategy. Knowledge Management Maturity (KMM) model was developed to address various forms of challenges and face challenges. . The concept of KMM model has been derived from Capability Maturity Model (CMM) of System Engineering and Integration (SEI). It serves a s two-fold-purpose. It provides a framework that can be used to access current level of Knowledge Management maturity. This framework also gives efforts on rising of level on knowledge management maturity. With the logic of CMM, the KMM model has generated five types of maturity levels. Key result areas are involved in each level. Each of the KRA defines a perfect capability in relation to people, processes, and technology. The company also focused on the content architecture. Various knowledge assets and types of knowledge have been incorporated under this juncture. Internal content of Infosys represents the internal expertise, i.e., project snapshot documents which creates a window into the project, reusable code, internal white papers, chat sessions, groups of discussions, and other artifacts. External content of Infosys represented outside expertise, i.e., reviewed websites, terms of technologies, online books and journals, technology summaries, external white papers, business news, and many more. Every item in the content repository must be associated with one or more than one node under knowledge hierarchy. In this way, it facilitates retrieval easily. The content under knowledge management repository has been charted into different stages, namely, streaming and editing, identification of internal experts, publishing, certification, and maintenance. The technology architecture is denoted by a central knowledge management portal. This helps to provide access to knowledge assets that have been defined by content architecture. In this field, a very important strategy is long-term technology vision which includes knowledge assets of all websites. These assets belong to different competency groups within the organization or to individuals in the central knowledge management architecture. The integrated accessibility becomes easy for the users. The organization, Infosys, has selected a special type of "facilitated approach", that has a forecast like (i) a central KM group has managed the technology architecture, (ii) all stages of content management procedure are anchored through central group, (iii) creation of internal content should be facilitated by the knowledge management group and must have proper practice on it. Another knowledge management initiative by Infosys is knowledge Currency Units or KCU. A knowledge currency unit scoreboard is attached with central KM portal. It will show high level of visibility toward strong contributors. This has an objective to establish correlation between making high profile organization and contribution toward knowledge management. Infosys conducts a "knowledge summit" every quarter. In this forum, contributors are facilitated and the concept of knowledge sharing must be highlighted. The process of knowledge sharing has been increased by quality processes. The top-level management helps to implement and maintain a perfect knowledge management strategy under Infosys functions. A steering committee (comprised of senior personnel) who formulate different high-level strategies and planning sessions on knowledge management paradigm. Members of the Infosys project team were distributed among customer sites as well as Infosys offshore software development centers on the globe. This distribution, combined with the rapidly changing technology landscape, posed a challenge for Infosys to update the knowledge and skills of its employees. Knowledge management group of Infosys, under its Education & Research Department, was charged with creating various assets and processes with the following objectives: (I) increasing reuse of knowledge, (II) facilitating higher functional effectiveness to enable competitive advantage, (III) using knowledge to improve quality continuously, (IV) leveraging knowledge assets to capture high-end business consulting engagements, and (V) enhancing the brand similar to universities, as to how they leverage knowledge-intensive research activities to do the same [41].

Access to Infosys KM systems usually is through Sparsh, corporate intranet. KShop (a KnowledgeShop) is the main knowledge management portal of Infosys used to connect to the company's computer knowledge warehouse. Besides this, Infosys offers a variety of other KM tools like Konnect (A professional networking platform to connect with people, discover expertise and share knowledge), Infyblogs (blogging tool), Team Wiki(tool to collaborate, share, and edit project knowledge with team members), KMail, Search + (searching for matching documents from a variety of sources such as Kshop, Konnect, Kmail), Discussion forum, etc. . Infosys has been very effective in leveraging past knowledge to become one of the world's leading IT service providers.

4.7.2 *Management at Motorola*

Information Technology industry is a nascent industry, its stages of evolution are continuous. It is a knowledge-based industry and deals with various forms of knowledge. The IT industry has become very much tech-savvy due to high (alt word) usage of technology. In earlier times, significance of knowledge and knowledge management was not so high due to the existence of technocrats who acted as the key decision-maker. Keeping in mind the importance of knowledge management system, we have focused and discussed on Motorola KMS, one of the recognized business organizations. Motorola is a multinational telecommunications company in the United States. From 2007 to 2009, they had suffered a loss of $4.3 billion and then the company split into two independent companies, namely Motorola Mobility and Motorola Solutions in the year 2011. It is considered that Motorola Solutions is the direct successor to Motorola Inc. After that Motorola Mobility has been accrued by the company Lenovo in the year 2014.

To maintain sustainable competition in the market, the IT industry has focused on knowledge and its recycling within the industry. Simply speaking, knowledge management in Motorola encompasses storage of information of customers and of various operations within the organization. The knowledge can be categorized further as data, information, and activities. Under the Motorola Company, the KM approaches are of two types, i.e., long term approach and short-term approach. The long-term approach is focused on growth of knowledge within the company when knowledge is organized in a continuous process. Therefore, management practices in relation to all the functional sections within the organization can be benefited from the use of knowledge management system. Big companies are following this long-term approach of KM in general because they already have a well-established knowledge management strategy. In this way, Motorola has a competitive advantage over other firms in the market. On the other hand, short-term approaches can be found in small-size companies. In Motorola, KM is implemented in the area of funding, requirement of human resources, customer support, decision making, problem-solving, etc.

The knowledge management system is emphasized in two ways within the company, (i) identification of need of KM which should be known to all and (ii) evolving the formal system for disseminating knowledge.

The KM system should be enabled in both flows under the juncture of communication process of business house, i.e., (i) a bottom-up information and (ii) a top-down information.

4.7.2.1 Knowledge Management Structure

At the initial level, knowledge management was under the premise of MIS (Management Information System) manager. His role was to collect data from different sources in the business. This was done at the level of operation, then the role was shifted to the high-level managerial stage and the position was denoted as a board member or Director MIS. At present knowledge experts or champions are performing across all functions. The role of these experts or champions is to coordinate KM activities at different levels. The various levels are functional level, semi-corporate level, and corporate level. The activities that are performed at each level are different from one another. Another parameter under KM in the said company is identification of knowledge requirements and dissemination of knowledge. The implementation of KM is very much focused on nature or characteristics of knowledge and then the methodologies which follow are involved in the process of dissemination of knowledge [42–45].

Under KM system, the implementation approaches are of two types, such as (i) Search-based approach and (ii) Hierarchy based approach. The search-based approach produces a superset of solutions from actual knowledge base. Like keyword search, this case is typically implemented in the system. This approach is totally search-engine oriented. By this approach, the users can evaluate each of the points within the space of resultant solutions. Then the best fit solution is opted. The disadvantage of this kind of approach is that the resultant solution space will be large if large size raw data exists. The benefit of KM system will then be reduced. In this case, the evaluation process is highly expensive and it involves extensive manpower involvement as well. On the other hand, the hierarchy-based approach of knowledge management provides a narrow subset of results. This hierarchy approach is facilitated by a tree-node approach. It compares current problems with existing ones in the knowledge base. It actually refines the parameters to identify the problem and finally it arrives with a narrow subset of solutions. The disadvantage of this approach is that it enables unidirectional solution to the problem and sometimes the solution might be forced to fit, though this is not the best way. In the real world, when colleagues and partners informally discuss a subject, always they try to arrive at a refined solution space. This phenomenon can be implemented in discussion-based approach which is better than the other two approaches, as large degree of customization is involved in the process. Knowledge management systems in Motorola should focus either on some kind of artificial intelligence system that can stimulate human interactions or should be enabled in real-time human interactions. Though it is very true that nowadays the implementation of AI-based systems is very expensive and complex to use as each point of problem space needs an algorithm, which must arrive at every point of solution space. The implementation tools used are web pages, project achieves, project post mortem reports, bulletin boards, and many more.

4.8 Conclusions

In this chapter, we have discussed the various aspects of knowledge management systems. At present, businesses have become very much competitive. Every business has a role to play that is goal-oriented. Every business organization wants to achieve its optimum level of profitability thereby leading to business growth and development. The various KMS software has made the journey easy for business firms. In this competitive era of business, when firms take advantage of an effective knowledge management system, they will be full of technological, manpower, and financial resources. All the internal employees and external customers both are getting facilities through KMS. In the strategic management of a business, it is essential to implement better knowledge management practices and norms that can provide the maximum business level output in less time. Nowadays, information technology is merged with different functional domains of management and makes all processes of work smooth. This possesses lower response time, lower cost, competitive advantage, and high production capacity. The business value and ethics can be generated promptly with the help of appropriate options from different categories of a knowledge management system. Therefore, to create national income in countries' assets, every business firm should be advanced, modern-methodological, and highly technical. The firm will create a positive image toward others and will influence others to adopt an efficient knowledge management system.

References

1. Maier, R., & Hadrich, T. (2011). Knowledge management systems. In *Encyclopedia of Knowledge Management* (pp. 779–790), Second Edn. IGI Global.
2. Becerra-Fernandez, I., & Sabherwal, R. (2014). *Knowledge management: Systems and processes*. Routledge.
3. Alavi, M., & Leidner, D. (1999). Knowledge management systems: Issues, challenges, and benefits. *Communications of the Association for Information systems, 1*(1), 7.
4. Thierauf, R. J. (1999). *Knowledge management systems for business*. Greenwood Publishing Group.
5. Edwards, J. S., Shaw, D., & Collier, P. M. (2005). Knowledge management systems: finding a way with technology. *Journal of Knowledge Management*.
6. Huber, G. P. (2001). Transfer of knowledge in knowledge management systems: Unexplored issues and suggested studies. *European Journal of Information Systems, 10*(2), 72–79.
7. Gallupe, B. (2001). Knowledge management systems: Surveying the landscape. *International Journal of Management Reviews, 3*(1), 61–77.
8. Meso, P., & Smith, R. (2000). A resource-based view of organizational knowledge management systems. *Journal of Knowledge Management*.
9. Wickramasinghe, N. (2003). Do we practice what we preach? Our knowledge management systems in practice are truly reflective of knowledge management systems in theory? *Business Process Management Journal*.
10. Carlsson, S. A. (2003). Knowledge managing and knowledge management systems in inter-organizational networks. *Knowledge and process management, 10*(3), 194–206.
11. Chen, H. (2001). Knowledge management systems: A text mining perspective. *Knowledge Computing Corporation*.

12. Adams, G. L., & Lamont, B. T. (2003). Knowledge management systems and developing sustainable competitive advantage. *Journal of Knowledge Management*.
13. O'Leary, D. E. (1998). Guest editor's introduction: Knowledge-management systems-converting and connecting. *IEEE Intelligent systems, 13*(03), 30–33.
14. Quaddus, M., & Xu, J. (2005). Adoption and diffusion of knowledge management systems: Field studies of factors and variables. *Knowledge-based systems, 18*(2–3), 107–115.
15. Hsia, T. L., Lin, L. M., Wu, J. H., Tsai, H. T., & Koohang, A. (2006). A framework for designing nursing knowledge management systems. *Interdisciplinary Journal of Information, Knowledge & Management, 1*.
16. Raman, M., Ryan, T., & Olfman, L. (2005). Designing knowledge management systems for teaching and learning with wiki technology. *Journal of information systems education, 16*(3), 311.
17. Benbya, H., & Belbaly, N. A. (2005). Mechanisms for knowledge management systems effectiveness: An exploratory analysis. *Knowledge and Process Management, 12*(3), 203–216.
18. Dorasamy, M., Raman, M., & Kaliannan, M. (2013). Knowledge management systems in support of disasters management: A two-decade review. *Technological Forecasting and Social Change, 80*(9), 1834–1853.
19. Bose, R. (2003). Knowledge management-enabled health care management systems: Capabilities, infrastructure, and decision-support. *Expert Systems with Applications, 24*(1), 59–71.
20. Young, M. L., Kuo, F. Y., & Myers, M. D. (2012). To share or not to share: A critical research perspective on knowledge management systems. *European Journal of Information Systems, 21*(5), 496–511.
21. Desouza, K. C. (2003). Barriers to effective use of knowledge management systems in software engineering. *Communications of the ACM, 46*(1), 99–101.
22. Halawi, L. A., McCarthy, R. V., & Aronson, J. E. (2008). An empirical investigation of knowledge management systems' success. *Journal of Computer Information systems, 48*(2), 121–135.
23. McCall, H., Arnold, V., & Sutton, S. G. (2008). Use of knowledge management systems and the impact on the acquisition of explicit knowledge. *Journal of Information Systems, 22*(2), 77–101.
24. Gottschalk, P. (2006). Stages of knowledge management systems in police investigations. *Knowledge-Based Systems, 19*(6), 381–387.
25. Wang, S., Noe, R. A., & Wang, Z. M. (2014). Motivating knowledge sharing in knowledge management systems: A quasi–field experiment. *Journal of Management, 40*(4), 978–1009.
26. Gao, F., Li, M., & Clarke, S. (2008). Knowledge, management, and knowledge management in business operations. *Journal of Knowledge Management*.
27. Davenport, T. H., & Glaser, J. (2002). Just-in-time delivery comes to knowledge management. *Harvard business review, 80*(7), 107–111.
28. Tseng, S. M. (2008). The effects of information technology on knowledge management systems. *Expert systems with applications, 35*(1–2), 150–160.
29. Dave, B., & Koskela, L. (2009). Collaborative knowledge management—A construction case study. *Automation in construction, 18*(7), 894–902.
30. Liebowitz, J. (2001). Knowledge management and its link to artificial intelligence. *Expert systems with applications, 20*(1), 1–6.
31. De Tienne, K. B., & Jackson, L. A. (2001). Knowledge management: understanding theory and developing strategy. *Competitiveness Review: An International Business Journal*.
32. Du Plessis, M. (2007). The role of knowledge management in innovation. *Journal of Knowledge Management*.
33. Borghoff, U. M., & Pareschi, R. (1997). Information technology for knowledge management. *Journal of universal computer science, 3*(8), 835–842.
34. Tatham, P., & Spens, K. (2011). Towards a humanitarian logistics knowledge management system. Disaster Prevention and Management: An International Journal.
35. Juang, Y. S., Lin, S. S., & Kao, H. P. (2008). A knowledge management system for series-parallel availability optimization and design. *Expert Systems with Applications, 34*(1), 181–193.

36. Revere, D., Turner, A. M., Madhavan, A., Rambo, N., Bugni, P. F., Kimball, A., & Fuller, S. S. (2007). Understanding the information needs of public health practitioners: A literature review to inform design of an interactive digital knowledge management system. *Journal of biomedical informatics, 40*(4), 410–421.
37. Durcikova, A., Fadel, K. J., Butler, B. S., & Galletta, D. F. (2011). Research note—knowledge exploration and exploitation: The impacts of psychological climate and knowledge management system access. *Information Systems Research, 22*(4), 855–866.
38. Liebowitz, J. (2004). A knowledge management strategy for the Jason organization: A case study. *Journal of Computer Information Systems, 44*(2), 1–5.
39. McInerney, C. (2002). Knowledge management and the dynamic nature of knowledge. *Journal of the American society for Information Science and Technology, 53*(12), 1009–1018.
40. Yeh, Y. J., Lai, S. Q., & Ho, C. T. (2006). Knowledge management enablers: A case study. *Industrial Management & Data Systems.*
41. Latha, A., Suresh, J.K., & Mahesh, K. (2010). KM in Projects: Methodology and Experience. In T. K. Srikantaiah, M. E. Koenig, & S. Al-Hawamdeh, (Eds.), *Convergence of project management and knowledge management* (pp. 145–173). Scarecrow Press. ISBN: 9780810876972
42. Laidlaw, F. J. (2003). Supporting internal technology transfer with knowledge management at Motorola: A case study. *International journal of technology transfer and commercialisation, 2*(1), 18–31.
43. Bureš, V., Tučník, P., Mikulecký, P., Mls, K., & Blecha, P. (2016). Application of ambient intelligence in educational institutions: Visions and architectures. *International Journal of Ambient Computing and Intelligence (IJACI), 7*(1), 94–120.
44. Singh, A., Sharma, A., & Dey, N. (2015). Semantics and agents oriented web personalization: state of the art. *International Journal of Service Science, Management, Engineering, and Technology (IJSSMET), 6*(2), 35–49.
45. Babo, R., Dey, N., & Ashour, A. S. (Eds.). (2020). Workgroups eAssessment: Planning, implementing and analysing frameworks (Vol. 199). Springer.

Chapter 5
Artificial Intelligence and Knowledge Management

5.1 Introduction

In recent years, data collection, segregation, and supervision plays a prime role in every enterprise. To achieve this, every enterprise spends large human resources and money to manage this information in an organized manner and this task is technically known as Knowledge Management (KM). The conventional methodology employed in KM helps to achieve the following operations in an organized manner: (i) Collection of the necessary information, (ii) Evaluating the collected information, (iii) Sharing the information within the enterprise or between enterprises, and (iv) Evaluating the available data to get the appropriate solution for a chosen task [1–4].

The traditional KM methodology employed in most of the enterprises is depicted in Fig. 5.1. The role of the KM is to maintain complete information about the necessary things which helps an organization to work in an efficient manner. Whenever a harsh situation arises, the organization will look for an optimal solution to tackle the situation and the existing KM system in the organization. The existing KM system also supports in getting all the necessary information which helps the organization to work in an optimal manner, which will help to achieve more profit by reducing the workload by getting the necessary expertise from an expert whenever it is necessary.

The various stages existing in the scheme are discussed below:

- **Vital information collection**: Collecting the necessary data during the day-to-day activities and creating a model based on the collected data will help to stimulate the organization's environment or to create a virtual organization model, which will help to identify and rectify the problems in the existing system. If the collected information is less, then it can be easily handled using conventional mathematical procedures. If the collected data is large (Big-Data), then a specific procedure is to be employed which will help to build the model of the organization and preserve the information for future use. This way the KM system is developed using the preserved information.
- **Reducing the duplication of efforts**: When a particular task is to be executed to perform work, time management is very essential. When a similar work is assigned

S. Majumder and N. Dey, *AI-empowered Knowledge Management*,
Studies in Big Data 107, https://doi.org/10.1007/978-981-19-0316-8_5

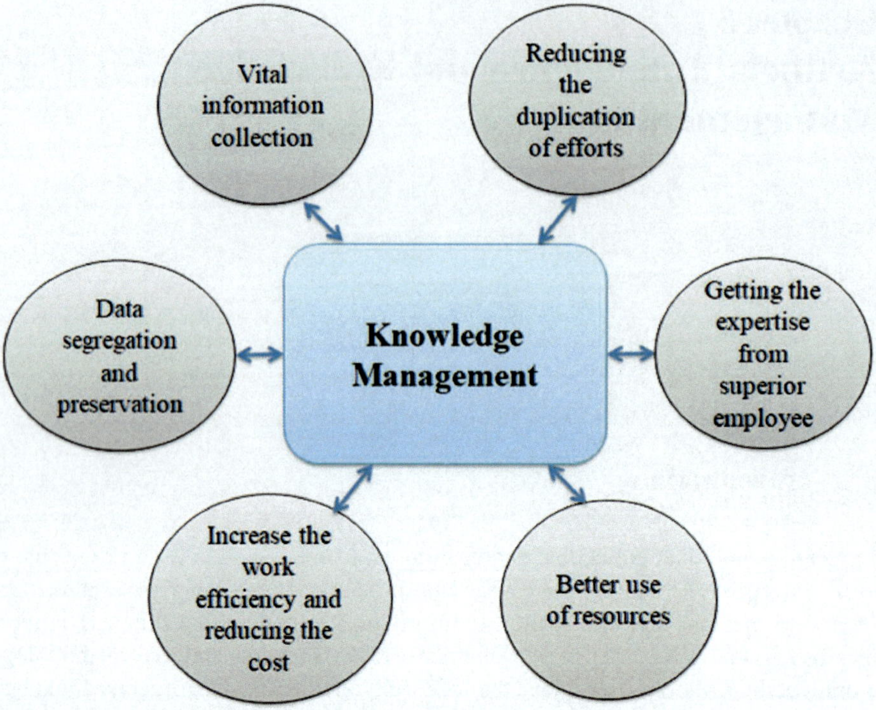

Fig. 5.1 Traditional KM scheme

to two different groups in the organization, the solution provided by one group can be verified and accepted by the other. This way, the time required to complete a single task will be reduced. When a KM system is existing in the organization, every group working on a problem is connected virtually, which helps to support the sharing of the knowledge and idea, which will reduce the duplication of the work and the task will be completed accurately within the specified time.

- **Getting the expertise from superior employee**: When the idea or the guidance from an expert member is recorded using an appropriate methodology, this will help when similar guidance is necessary in the future. The stored expertise will also support speedy decision making during harsh situations. The stored expert information also helps to get better high-quality outcomes and this information can be shared with other organizations, if necessary and the existing information can be altered based on the need.
- **Better use of resources**: Every organization's growth and its profits depends mainly on utilizing the resource in an efficient way. In every organization, the better utilization of the resources depends mainly on the support of staff, and improving the knowledge of the staff is very essential to have a better profit. The employee empowerment process can be done with scheduled training as well as self-help. When the organization is having a better KM system, then employee empowerment will be achieved easily which will help to utilize all the necessary

resources in an efficient manner. The role of the KM system is very essential to educate the staff in the organization about the handling of the resources.

- **Increase the work efficiency and reducing the cost**: When a well-trained staff is existing, then their work efficiency is good and this will help the organization to achieve a better profit. When every employee is perfectly trained, then decision-making capability during a task will be improved it will help to get the best possible work efficiency to complete a specified task. This optimal work will also reduce the utilization of the resources and improve the overall profit of the industry.
- **Data segregation and preservation**: Data collection is a vital process to complete examine a situation in an organization and the collected data will provide the necessary information regarding the work to be executed. When the data is effectively collected and segregated, then it will help to create a model which will help to predict the outcome. The data preservation also will support in getting the necessary information for the future examination of the collected data.

5.1.1 Information Collection for Efficient Decision Making

In every organization, data collection and preservation helps to build an efficient KM system, which supports the necessary decision-making process, whenever it is essential. Based on the volume of the data, it can be categorized as conventional and big-data, and the collection, pre-processing, and preservation procedure varies based on the volume of the data. Figures 5.2 and 5.3 depict the data collection procedures employed for conventional and big-data, respectively.

Figure 5.2 depicts the different phases existing in the simple data-supported KM scheme and it involves data collection, information processing, knowledge extraction

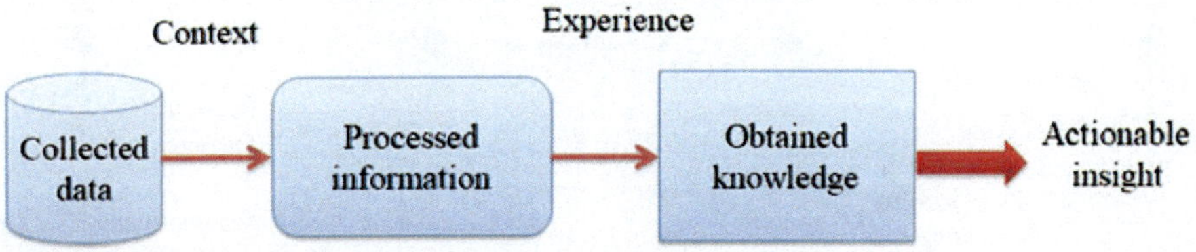

Fig. 5.2 Conventional data management

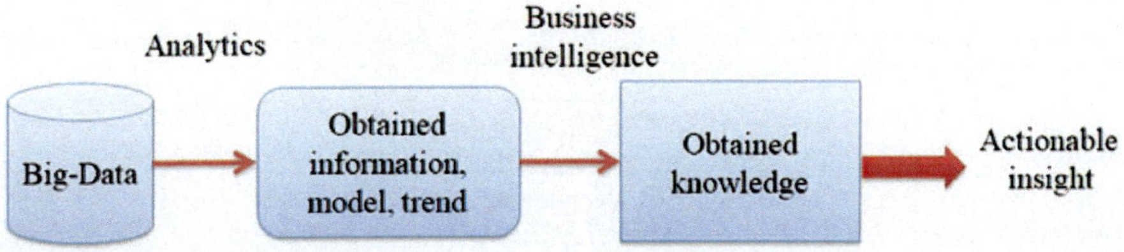

Fig. 5.3 Big-data management

from processed information, and building the KM. In this scheme, the data along with the context helps to get the information, and based on the expertise, the necessary knowledge will be extracted which helps to develop an actionable insight.

Figure 5.3 shows the different stages involved in big-data-supported KM development. After collecting the necessary data, a recommended examination process is implemented to achieve the information, model created with data, and its trend. Then the computer-supported scheme is implemented to extract the knowledge to get the insight. The processing procedure involved in big data is very complex compared to the traditional data and hence, the Artificial Intelligence supported KM scheme (AIKM) is proposed and implemented in various organizations to simplify the resource handling as well as decision-making tasks.

The conventional AIKM scheme widely found in various enterprises can be found in Fig. 5.4.

As discussed earlier, the AIKM system [5–8] is widely employed in organizations, when the necessary information/knowledge is to be extracted and preserved from the big data. The conventional data assessment procedures need complex data handling procedures when the data volume is large. Further, the processing time taken by most of the traditional knowledge extraction schemes employed for the big data is more, hence the AIKM systems are developed and employed in most organizations. Further, in AIKM, a chosen AI will monitor and control the whole operation as per the priority, which will help to get a better decision-making capability compared to the traditional techniques. The AIKM consists the phases, such as conventional

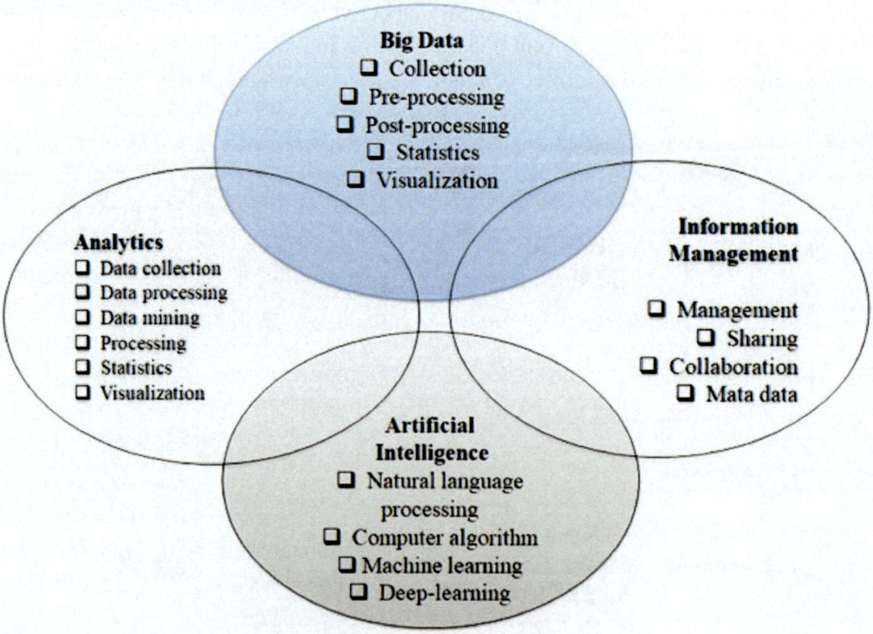

Fig. 5.4 Conventional artificial intelligence supported KM system

data analytics section, big-data collection, and processing, chosen AI schemes [9] to provide the solution, and information management.

The data collection phase helps to gather vital information from various parts of the organization and segregates the data based on its nature. The data may be a simple signal, image, resource sample information, expenditure, production cost, profit, etc. After collecting the information as per the schedule/priority, it will segregate the data in order for further processing. This section also executes the tasks, such as visualization of the data, pre-processing the data, extracting the vital information using a chosen data mining technique, and mathematically evaluating the data to create the model. This procedure is common for both the conventional and big data and this helps to collect and convert the information into a usable form. After treating the raw data using a chosen technique, the AI schemes are then employed for optimization, implementing the Natural Language Processing (NLP) procedures to extract the knowledge, employing the machine/deep learning system for data processing and knowledge extraction, etc. After extracting the necessary information, all these information are preserved using a KM system. The information management system in AIKM aims to perform, data preservation, sharing the data to the user based on the need, collaborative decision-making support, and developing the meta-data by accumulating the best possible data provided by the AI.

From the above information, it can be confirmed that the big-data-based KM scheme needs the support of the AI to get the finest outcome whenever is necessary. The impact of AI can be.

5.2 Impact of Artificial Intelligence in Knowledge Management

The prime need of KM is to permit the user and enterprise.to work together, develop, distribute, and use the information (knowledge) collectively [10–14]. When an organization is equipped with the necessary KM system, then it will support performance improvement, supports modernization, and help to get the essential awareness about the various process involved in the organization. The preserved information must be active, truthful, and individual to be considered during the decision-making tasks. In AIKM, the machines (computer/processor) are permitted to learn from the existing as well as the current data to find the best possible information that can be transferred to the staff of the organization to enhance the decision-making process. AI also supports digitizing the entire data to support faster and more accurate knowledge delivery, whenever it is needed. In most cases, AI provides the necessary information to the people to make the right decision, whenever is needed. In some situations, AI itself provides the necessary decision, which can be verified and accepted by the people, who belong to the organization. From this, it is clear that the AI-based KM helps the people to make the right decision or the AI itself provides an unbiased decision based on the existing data.

In recent years, the AIKM is employed in every organization to support the necessary data collection, data preservation, and decision making. The implemented AIKM's structure varies based on the organization.

For example, the employed AIKM in the healthcare domain is different compared to other domains, such as financial sectors, engineering, and industry. In the healthcare domain, the AIKM is employed to preserve the necessary information about the hospital, medicines available, doctors, and patients. The scheme employed in this sector is unique and it cannot be the same compared to the AIKM employed in other organizations.

Every modern hospital is equipped with a sufficient facility to employ the AIKM, which involves disease-related data collection, data processing, model development or statistical evaluation of processed information, knowledge mining, decision making, and treatment. When the AI schemes are employed in the healthcare domain, the patient who is admitted to the particular hospital will get the essential benefits, such as accurate diagnosis, early detection of the disease and its harshness, obtaining the necessary preventive services to control the spread of the disease, enhancing the clinical level decision making, determining latest handling as well as medication facilities, providing personalized healthcare. Further, the AI-based KM system also helps to monitor the inhouse/remote patients with the latest technologies using the computerized monitoring tools interconnected using the dedicated communication network. In addition to that, the AIKM helps to hide personal data, such as patient information, disease condition, treatment implemented, and other information which will affect the patient's reputation. Every information is made available only when a request is raised by the patient or the authorized member related to the patient.

The AIKM employed in hospitals helps the patient to get the necessary information based on the request. Further, the healthcare division is a knowledge-intensive domain and always monitors and supports the necessary data collection and monitoring in order to confirm the right treatment to the patient based on the need. The AIKM also helps to develop and maintain the Electronic Health Records (EHR) for every disease, which helps to monitor and treat the other patient having similar disease symptoms.

The AIKM supports the complete automation of the hospital, which enhanced the modern hospitals by employing the necessary computer software, which schedules the medication process for every patient and supports the complete monitoring of patients with the help of suitable monitoring equipment. The recent works in the literature confirm that AI is widely adopted in modern hospitals to automate, disease diagnosis, decision making, treatment execution, and recovery monitoring. The conventional structure of the AI-supported patient monitoring system is depicted in Fig. 5.5. This figure authenticates that, the AI-based techniques, helps to achieve:

(i) Premature disease detection: Every modern hospital is equipped with AI technique based disease detection. Normally, the monitoring computer employed in the AI method is already trained with a considerable number of disease data and this scheme is also associated with the necessary disease models. When

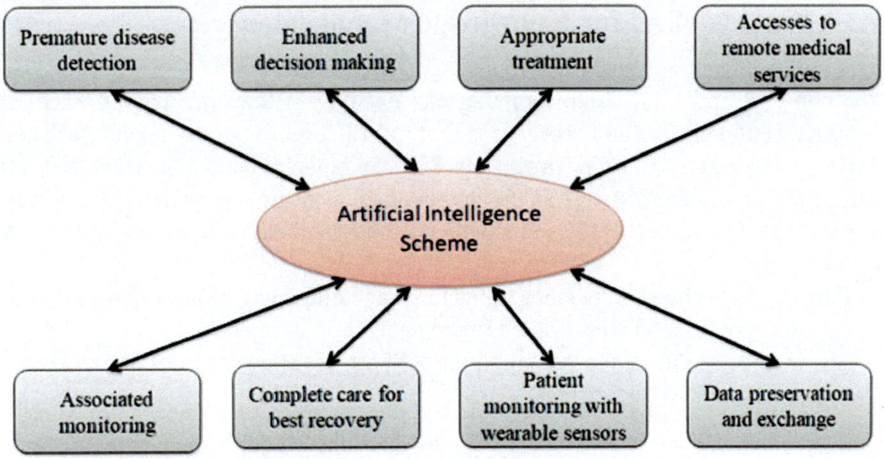

Fig. 5.5 Artificial intelligence supported patient monitoring system

a new patient is admitted for the treatment, the necessary disease information is then collected from the patient and this information is then compared against the existing data/model in the AI to get a faster disease prediction and treatment execution. The AI also collects the following information from the patient: disease symptom, its harshness, treatment procedures, medication to be implemented, and recovery time, and the collected data is then considered to get faster disease detection when compared to a traditional scheme. The existing AI system is already trained with the necessary disease information and when it received the collected information/symptoms from the patient immediately it will predict the disease and instruct the doctor regarding the other tests to be performed to confirm the disease. The AI-supported premature disease diagnosis will help to identify a class of infectious diseases and acute diseases in its early phase and this will avoid the delay in treatment.

(ii) Enhanced decision making: Most of the hospitals in low/middle-income countries are having a lesser value of doctor to patient ratio and hence a considerable delay is happening in the disease diagnosis and decision-making process. To avoid this limitation, most of the recent hospitals are equipped with automated disease detection systems, which will act as the supporting tool for doctors. The initial level of the disease detection and decision making is done by the computer algorithm and the generated report is then sent to the doctor for further verification and approval. The various stages involved in this pipeline include: (i) Collection of the necessary information from the patient, (ii) Collection of the clinical information, (iii) Comparing the collected information with the similar existing information/model in the AI, (iv) Confirming the disease and generating the report for further analysis, (v) Getting the approval from the doctor and treatment implementation.

5.3 Benefits of AI for Knowledge Management

The complex and high volume of the data existing in the organization needs to be processed to extract and preserve useful information for future needs. When an AIKM is implemented, every process in KM can be automated and which helps to reduce the processing as well as the data management time. Further, the AI helps to get timely and accurate decisions, which cannot be done with the traditional KM scheme [15].

Further, the AI helps to balance the volume and efficiency of information sharing among the people, who are involved in the decision-making tasks.

The major benefits of the AI-supported KM are as follows:

- Develops a model based on the collected data to predict the current areas/topics related to awareness, which every employee of the organization needs.
- AI helps to recognize the targeted information which targets every employee of the organization based on real-time commitment and expenditure.
- It supports the computerized processing as well as personalize the information based on employee's choice.
- The implemented AI schemes, such as machine/deep learning techniques help to enhance the content decisions which provides the optimal and precise solution based on the situation.
- It initiates a preferred or auto search in KM system to get the applicable, accurate, and resourceful information to support the right decision making based on the need.
- The AI integrated with the NLP will provide the necessary support for each employee working in different organizations, who needs assistance during the critical decision-making tasks.
- In some special cases, the AI itself makes the right and unbiased decision based on the problem to be solved. The AI-assisted models and the existing statistics will also help to get better decisions.

5.4 Future of Knowledge Management

The KM system is very essential in every organization for data handling and decision making. The precisely implemented KM scheme helps to get a better profit and optimized resource utilization [16–20]. The future of the KM system includes:

- ***Use of Social Networks***: In the current era, Social Network (SN) plays a vital role in connecting a variety of people together to share their idea about a particular/general problem. The digital computing atmosphere is rapidly growing due to the various advancements happening in the Computer and Information

Technology (CIT) domain. The recent advancement in CIT helped to integrate the people/employees under a common/diverse software-supported platform. To support this facility, a considerable number of mobile/computer applications (APP) are developed and which links the people/employee, who get the membership in the APP with a secured registration.

When a group of people/employees is connected through a secured APP-supported link, the idea, as well as the solution, can be easily shared among the group which will reduce the decision-making time when a problem arises. Further, the availability of SN also helps to collect sufficient data from the group members, which can be treated using a chosen AI technique to extract and preserve the information/knowledge. The preserved knowledge can be considered as the approved Electronic Record (ER) for a chosen problem and when a similar problem arises, the necessary solution can be obtained by just looking at the previous ER of the similar problem. The existing ER can also be considered as a resource for training the people/employee of the organization.

In the future, the necessary SN APP can be created for a particular organization through which the employees can be integrated in order to participate in (i) Idea discussion, (ii) Transferring the knowledge from an experienced employee to the other person, (iii) Sharing and correcting the preserved information, (iv) Faster decision making to improve the outcome of the organization and (v) Connecting one organization with another through a dedicated communication link to support the knowledge sharing, etc.

Figure 5.6 depicts the structure of the SN-based KM system. In every organization, knowledge sharing is an essential process in daily tasks among employees, skilled professionals, and other sections, which needs to be monitored efficiently. Further, in every organization, the preserved information/knowledge is considered to be a prime capital asset and it must be shared to the necessary location efficiently during strategic plans and decision making. When this process is implemented using the recent advancements, such as SN-based APP, big-data analytics, secured data preservation using a dedicated server as well as the cloud server. When the suggested method is employed, which helps the organization to share the information within the group and also organization to organization.

- ***Implementation of various collaborations to exchange data***: In KM scheme, the data collection from the right location plays a vital role to obtain the necessary information/knowledge for data preservation and updation. The availability of various communication facilities improved this task. Further, a considerable number of software-driven system and user-friendly APP helps to exchange the data from one location to other with employed security. When collaborative knowledge sharing is executed with a chosen scheme, which will help the organization to get various opinions from various levels of employees, such as skilled labor, working employees, and the domain expert. The necessary data is collected from various levels of the organization, it is processed and segregated based on its level and preserved. The preserved data is then exchanged among the levels for further

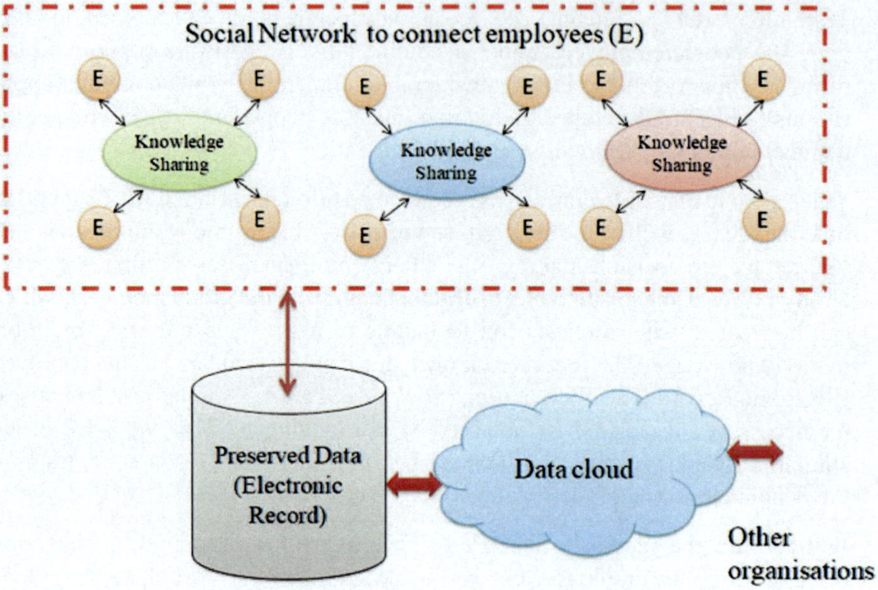

Fig. 5.6 Social network supported KM system

improvement or training purposes. The finest data selection and its exchange among the group members can be achieved using a chosen AI scheme, which considerably reduces the data handling time.

- ***AI-based Search and Data Handling***: AI scheme-based data handling and KM is sufficiently discussed in the earlier section and the architecture of this technique is presented in Fig. 5.4. The chief merit of AI-based data handling is it supports the automated technique to enhance the various phases of KM, such as data collection, processing, information extraction, and knowledge preservation. In this, a chosen AI scheme (machine-learning, deep-learning, and Neural Network based system) is employed to automate the whole KM system. The AI-supported scheme is used to achieve the following task: (i) Assisting the employee/user to make the decision for a chosen problem and (ii) Provide a computer-generated decision to support the faster decision making. In the future, the AIKM system can be developed and connected to the cloud for better data handling and preservation.
- ***Distributed cloud and Hybrid cloud***: The recently developed Cloud and Fog computing facility considerably reduced the data handling process and supports [21–24] efficient as well as quick data exchange, whenever a request is raised. The main use of this facility is to replace the traditional standalone data server with the recently developed techniques. When a cloud environment is employed in organization to support the data exchange, it helps to support the data access from various authorized remote locations. Based on the structure, this computing scheme is categorized into a distributed cloud environment and a hybrid cloud environment. In both schemes, the data exchange is to be done between the organization and the

cloud server. The server receives the request, process the request, and transfers the knowledge, which helps to get a better decision during a problem-solving task. Figure 5.7 presents the scheme of the distributed cloud system in which the autonomous decision-making system will interact with the cloud through a shared computing system, information unit, and pervasive system. This scheme normally helps to get all the necessary information from the earlier data and helps to find the optimal as well as an autonomous solution for the chosen task.

The hybrid cloud shown in Fig. 5.8 is the combination of the public and private cloud in which the tab sharing from the user to the cloud is performed using a controlled/uncontrolled manner. Normally, the organization will communicate with the cloud through a secured gateway and the public/general employee will have the conversation with an unsecured gateway. The cloud collects all the necessary information and shares it with the KM scheme based on the need.

- ***Hyperautomation in Knowledge Management***: Automation in organization is normally achieved by combining the necessary software with the associated hardware section. The automated scheme will help to support computer-based handling of the data which provides the necessary insight about the problem to be solved in an efficient manner. The various sections existing in this scheme are depicted

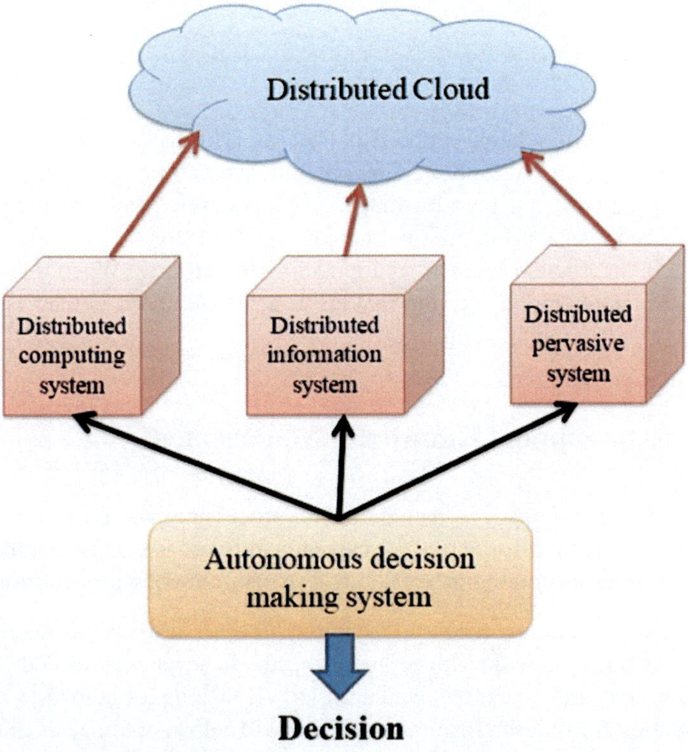

Fig. 5.7 Distributed cloud for data exchange

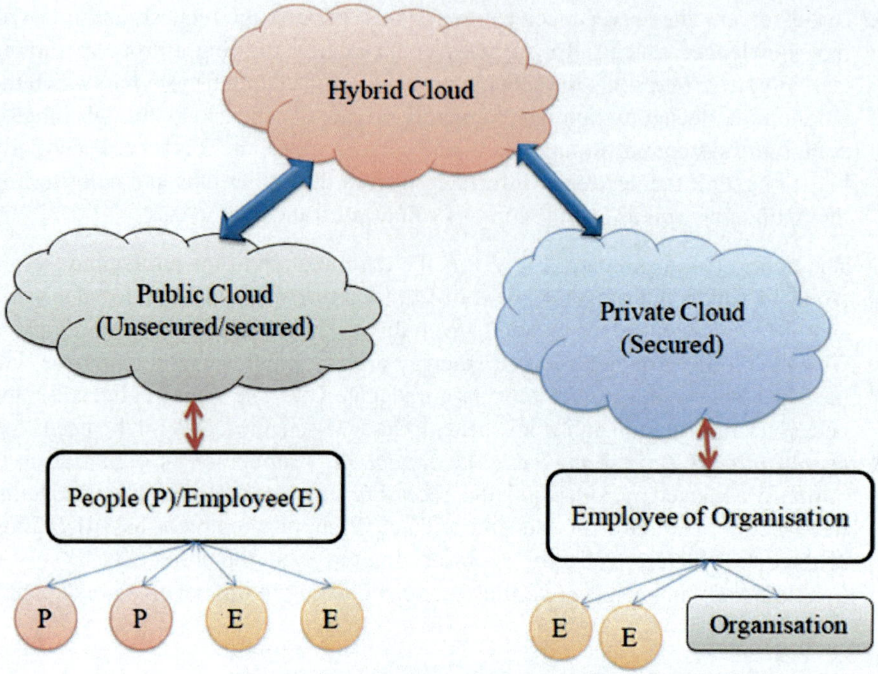

Fig. 5.8 Hybrid cloud to support the data sharing from various levels

in Fig. 5.9 and this scheme supports the necessary decision making based on the developed computer algorithm. This scheme involves various AI schemes such as Neural Network, machine-learning, and deep-learning techniques for handling the collected data. Further, it also includes the NLP for task mining, analyzing the data in an efficient manner, and modeling the process. When this scheme is employed, autonomous decision making is achieved and it is then verified and approved by the organization.

5.5 Tools to Support Knowledge Management

In the current era, a number of data handling, processing, and preserving tools are developed to maintain information for current and future use. Efficient information (knowledge) creating and maintenance involves the following procedures:

- Information development: This is the initial process, in which the necessary data for the organization is developed and utilized whenever it is needed or as per a schedule. In recent years, the availability of AI techniques helped to reduce the data creation and distribution process, which will help to optimize the information which is shared by the company to employees.

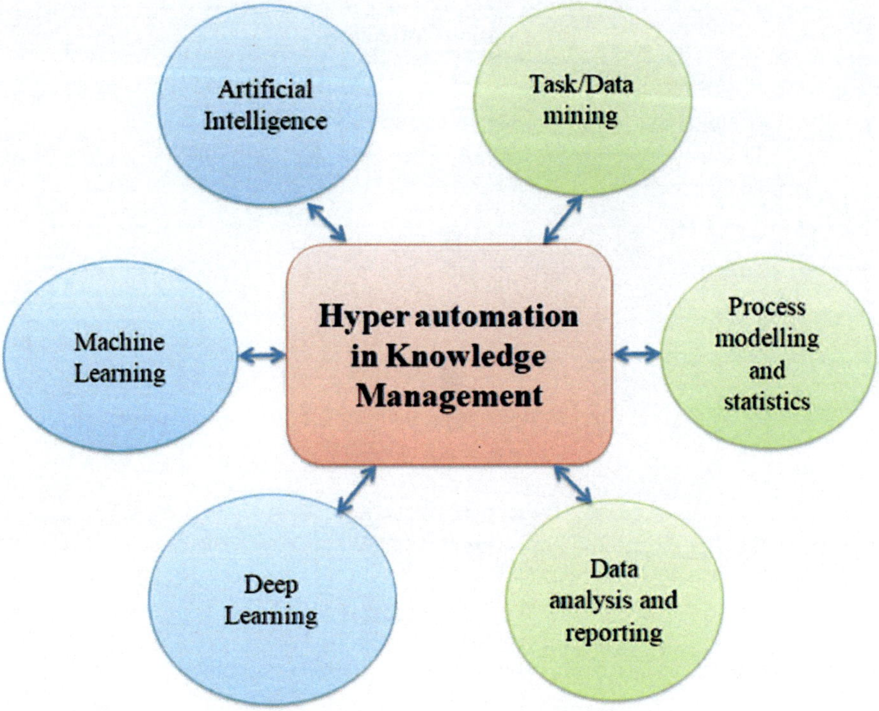

Fig. 5.9 Block diagram of hyperautomation in knowledge management

- Information preservation: This is the prime part of the knowledge handling process, which is considered a dedicated information technology scheme to create the managerial level information for the distribution process. The stored data/information must be processed in a particular technique in order to satisfy the requirements of the organization's warehouse.
- Information Sharing: The available information is to be communicated largely across the institute. The speed at which information reaches the levels of the employee will differ based on the organizational background. Companies that support and reward this performance will have a considerable number of merits and this process also will help to build a complete communication system through which the necessary information can be shared based on the need.

Figure 5.10 depicts the various procedures involved in information collection, handling, and preservation.

To support this process, a number of customary software and hardware systems are utilized. The major tasks performed by this system includes:

- Document managing scheme which works as a central storage structure for necessary digital documents. This scheme improves employee workflows by assisting the simple recovery of documents.

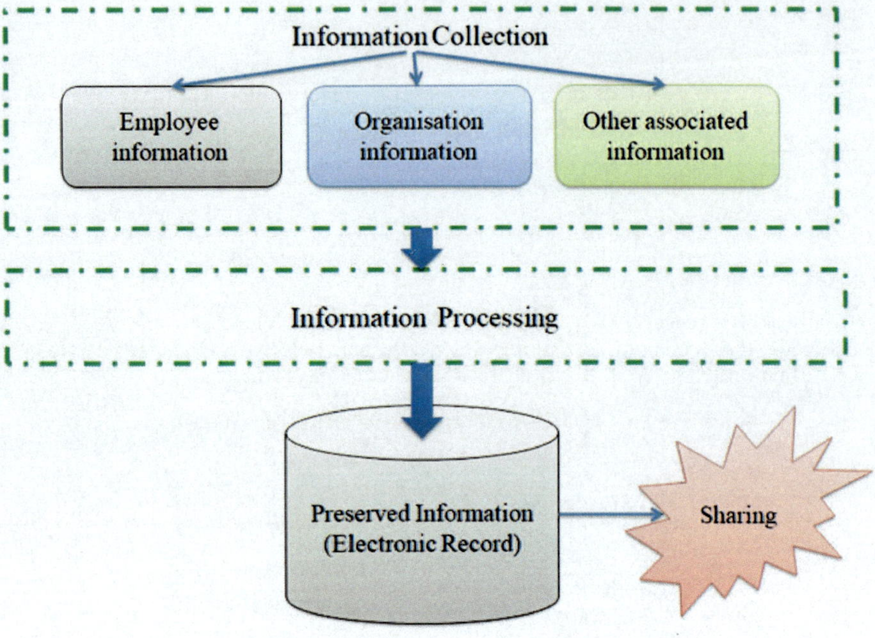

Fig. 5.10 Information management in organization

- Content organization structure (COS) are methodologies to handle the online/web-related information to support the users to handle the complete information (modify, share, and reshare the information). Along with the test information, the COS also supports the handling of complex information, which are in the form of images, signals, and video.
- Recently, every organization and its regions are connected using the necessary communication link. These networks help to exchange knowledge from one region to other with considerable accuracy. The traditional methods of knowledge sharing are more time-consuming and the improved communication facilities support enhanced sharing, modification of the shared information, and discarding the shared information (if necessary). This scheme helps to share the individual as well as the collaborative knowledge sharing.
- In recent years, every organization is equipped with the necessary data processing software and third-party support to collect, process, and preserve the data. Further, the recent advancements, such as AI techniques, IoT, and Cloud [25–29] helped to form a wider knowledge-sharing network, through the in-house and remote knowledge sharing can be achieved. Further, when this system is associated with a feedback scheme, the employee can help to convey the information regarding the knowledge which he/she received, which will ensure the appropriate sharing of information within and in between organizations [30, 31].

5.6 Conclusions

Information management is an essential task in every organization in order to predict its growth and to set future goals. Knowledge Management (KM) is an essential process in the organization, which helps to guide/train new employees based on the need. The KM employed in earlier days is a complex and time-consuming process and dedicated employees are appointed to support, the collection of information, processing, preserving, and sharing the knowledge, based on the need. In recent years, the progress in AI schemes helped the KM to its next level. The recent advancements, such as information optimization, IoT, and cloud enhanced the KM systems existing in the industries. When the AI-based KM is employed, handling of the information sharing process is automated, which will reduce the data handling burden that existed in the earlier KM scheme. Further, the AI-based schemes help to support the accurate and timely knowledge sharing process, based on the need.

References

1. Liebowitz, J. (2001). Knowledge management and its link to artificial intelligence. *Expert Systems with Applications, 20*(1), 1–6.
2. Tsui, E., Garner, B. J., & Staab, S. (2000). The role of artificial intelligence in knowledge management. *Knowledge Based Systems, 13*(5), 235–239.
3. Nemati, H. R., Steiger, D. M., Iyer, L. S., & Herschel, R. T. (2002). Knowledge warehouse: An architectural integration of knowledge management, decision support, artificial intelligence and data warehousing. *Decision Support Systems, 33*(2), 143–161.
4. Metaxiotis, K., Ergazakis, K., Samouilidis, E., & Psarras, J. (2003). Decision support through knowledge management: the role of the artificial intelligence. *Information Management and Computer Security*.
5. Becerra-Fernandez, I. (2000). The role of artificial intelligence technologies in the implementation of people-finder knowledge management systems. *Knowledge-Based Systems, 13*(5), 315–320.
6. Gacanin, H. (2019). Autonomous wireless systems with artificial intelligence: A knowledge management perspective. *IEEE Vehicular Technology Magazine, 14*(3), 51–59.
7. Sanzogni, L., Guzman, G., & Busch, P. (2017). Artificial intelligence and knowledge management: Questioning the tacit dimension. *Prometheus, 35*(1), 37–56.
8. Gandon, F. (2002). *Distributed artificial intelligence and knowledge management: Ontologies and multi-agent systems for a corporate semantic web* [Doctoral dissertation, Université Nice Sophia Antipolis].
9. Furmankiewicz, M., Sołtysik-Piorunkiewicz, A., & Ziuziański, P. (2014, July). Artificial intelligence systems for knowledge management in e-health: The study of intelligent software agents. In *Latest Trends on Systems: The Proceedings of 18th International Conference on Systems* (pp. 551–556). Santorini Island, Greece.
10. Liao, S. H. (2003). Knowledge management technologies and applications—Literature review from 1995 to 2002. *Expert Systems with Applications, 25*(2), 155–164.
11. Hoeschl, H. C., & Barcellos, V. (2006, August). Artificial intelligence and knowledge management. In *IFIP International Conference on Artificial Intelligence in Theory and Practice* (pp. 11–19). Springer.
12. Ziuziański, P., Furmankiewicz, M., & Sołtysik-Piorunkiewicz, A. (2014). E-health artificial intelligence system implementation: Case study of knowledge management dashboard of

epidemiological data in Poland. *International Journal of Biology and Biomedical Engineering, 8*(1), 164–171.
13. Paschek, D., Mocan, A., Dufour, C. M., & Draghici, A. (2017). Organizational knowledge management with Big Data. The foundation of using artificial intelligence. *Balkan Region Conference on Engineering and Business Education, 2*(1), 301–308.
14. Avdeenko, T. V., Makarova, E. S., & Klavsuts, I. L. (2016, October). Artificial intelligence support of knowledge transformation in knowledge management systems. In *2016 13th International Scientific-Technical Conference on Actual Problems of Electronics Instrument Engineering (APEIE)* (Vol. 3, pp. 195–201). IEEE.
15. Metaxiotis, K., Ergazakis, K., Samouilidis, E., & Psarras, J. (2004). Decision support through knowledge management: The role of the artificial intelligence. *International Journal of Computer Applications in Technology, 19*(2), 101–106.
16. Wiig, K. M. (1999). What future knowledge management users may expect. *Journal of Knowledge Management*.
17. Barclay, R. O., & Murray, P. C. (1997). What is knowledge management. *Knowledge Praxis, 19*(1), 1–10.
18. O'Leary, D. E. (1998). Enterprise knowledge management. *Computer, 31*(3), 54–61.
19. Becerra-Fernandez, I. (2004). *Knowledge management: Challenges, solutions, and technologies.*
20. O'Leary, D. E. (1998). Using AI in knowledge management: Knowledge bases and ontologies. *IEEE Intelligent Systems and Their Applications, 13*(3), 34–39.
21. Khoshnevis, S., & Rabeifar, F. (2012). Toward knowledge management as a service in cloud-based environments. *International Journal of Mechatronics, Electrical and Computer Technology, 2*(4), 88–110.
22. Abdullah, R., Eri, Z. D., & Talib, A. M. (2011, November). A model of knowledge management system for facilitating knowledge as a service (KaaS) in cloud computing environment. In *2011 International Conference on Research and Innovation in Information Systems* (pp. 1–4). IEEE.
23. Aksoy, M. S., & Algawiaz, D. (2014). Knowledge management in the cloud: Benefits and risks. *International Journal of Computer Applications Technology and Research, 3*(11), 718–720.
24. Abdullah, R., & Alsharaei, Y. A. (2016, August). A Mobile Knowledge as a service (mKaaS) model of knowledge management system in facilitating knowledge sharing of cloud education community environment. In *2016 Third International Conference on Information Retrieval and Knowledge Management (CAMP)* (pp. 143–148). IEEE.
25. Goswami, S., Roy, P., Dey, N., & Chakraborty, S. (2016). Wireless body area networks combined with mobile cloud computing in healthcare: A survey. In *Classification and clustering in biomedical signal processing* (pp. 388–402). IGI Global.
26. Matallah, H., Belalem, G., & Bouamrane, K. (2017). Towards a new model of storage and access to data in big data and cloud computing. *International Journal of Ambient Computing and Intelligence (IJACI), 8*(4), 31–44.
27. Sarkar, M., Banerjee, S., Badr, Y., & Sangaiah, A. K. (2017). Configuring a trusted cloud service model for smart city exploration using hybrid intelligence. *International Journal of Ambient Computing and Intelligence (IJACI), 8*(3), 1–21.
28. Dey, N., Hassanien, A. E., Bhatt, C., Ashour, A., & Satapathy, S. C. (Eds.). (2018). *Internet of things and big data analytics toward next-generation intelligence* (Vol. 35). Springer.
29. Gupta, N., Gupta, S., Khosravy, M., Dey, N., Joshi, N., Crespo, R. G., & Patel, N. (2021). Economic IoT strategy: The future technology for health monitoring and diagnostic of agriculture vehicles. *Journal of Intelligent Manufacturing, 32*(4), 1117–1128.
30. Gupta, N., Khosravy, M., Patel, N., Dey, N., Gupta, S., Darbari, H., & Crespo, R. G. (2020). Economic data analytic AI technique on IoT edge devices for health monitoring of agriculture machines. *Applied Intelligence, 50*(11), 3990–4016.
31. Elhayatmy, G., Dey, N., & Ashour, A. S. (2018). Internet of Things based wireless body area network in healthcare. In *Internet of things and big data analytics toward next-generation intelligence* (pp. 3–20). Springer.

Chapter 6
Explainable Artificial Intelligence (XAI) for Knowledge Management (KM)

6.1 Introduction

Due to access to large datasets and advances in computing power, machine learning models for prediction and analysis problems have gained significant importance since the last decade [1]. The methods of ML are widely used to develop decision systems for disease detection and drug prescription. We know that AI is very powerful and provides enormously accurate results. Yet it is extremely opaque, and accuracy should not be the only measure of trust. We can also say that it is like a black box, where the outside observer does not know what is happening inside the box [2]. It becomes highly risky and dubious when we entrust such important decisions to a system that cannot tell us how it came to that conclusion. The difficulty for the system in not being able to explain its answers is also called the black box problem. Science and society are far from being concerned only with performance. The problem widens as concerns about ethical AI increase. It is difficult to explain the details of such black box models to a person with a different background.

Explainable Artificial Intelligence (XAI) is a transparent window that displays internal processing in machine learning models. Figure 6.1 shows what the main differences are between XAI and AI. Explainability in this case means understanding how a model works, but not in such a complex way that it is difficult for an average person to understand [3–5]. Explainability tailors explanations to the needs of the end user and their comprehension skills.

6.2 Need of XAI

The overall goal of XAI is to help people understand and rely on the results of Artificially Intelligent or Machine Learning. XAI is an attempt to make these AI/ML models more transparent without compromising their performance. This in turn allows people to understand how these models actually work. Trust is an important

S. Majumder and N. Dey, *AI-empowered Knowledge Management*,
Studies in Big Data 107, https://doi.org/10.1007/978-981-19-0316-8_6

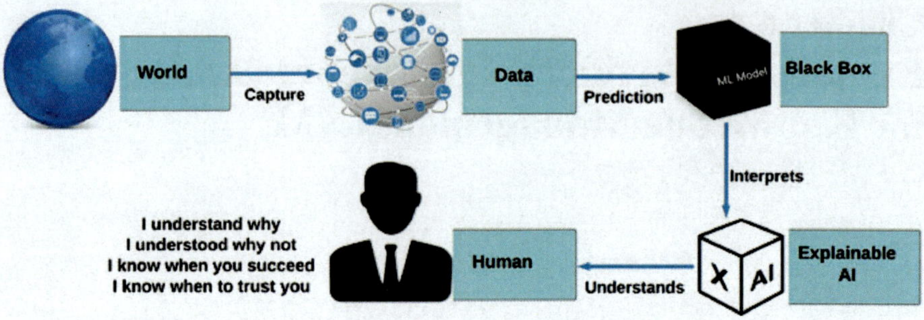

Fig. 6.1 Workflow of explainable AI

factor when developing or using such models to make important decisions. Users or owners may want to know why the model decides what it decides, how the model comes to a conclusion. So to use a model by law and regulation, you should be able to show exactly how it works.

Transparency in these algorithms will help ensure that only meaningful variables determine the results and that the model is not biased. Transparency will also help in the evolution of the model, probably by correcting errors in the system or developing new theories. The term "explainable" was used because the focus was on human understanding. The main interest was in the human psychology of explanation. The explanation should include what information the model uses to make decisions whether it understands how things work and what its goals are. XAI brings many benefits to scientists, industrialists, and the workers who use the system. XAI will simply make things easier to interpret. XAI will help build user confidence in the system or model, meet any legal requirements, and provide ethical justification because we will be able to understand how the system really works. XAI will provide a better understanding and interpretation of the results. XAI is evolving into a new method to help people better understand how a model makes decisions and comes to its conclusions. Understanding how a system works is of great importance in today's world. XAI will be needed by many industries as it will help understand the insights and predictions generated by the AI or ML models in their systems. XAI will help achieve the results that industries and businesses value: high efficiency and low cost.

6.3 Taxonomy of XAI

A variety of taxonomies have been proposed to classify explanatory strategies. There is no universal classification system for explanatory strategies. They can vary widely depending on the characteristics of the methods and are classified into numerous overlapping or non-overlapping classes. The taxonomy is described in Fig. 6.2.

1. Local method: the basic idea of a local interpretable model is to explain a single prediction by focusing on a single event and analyzing how the model arrived

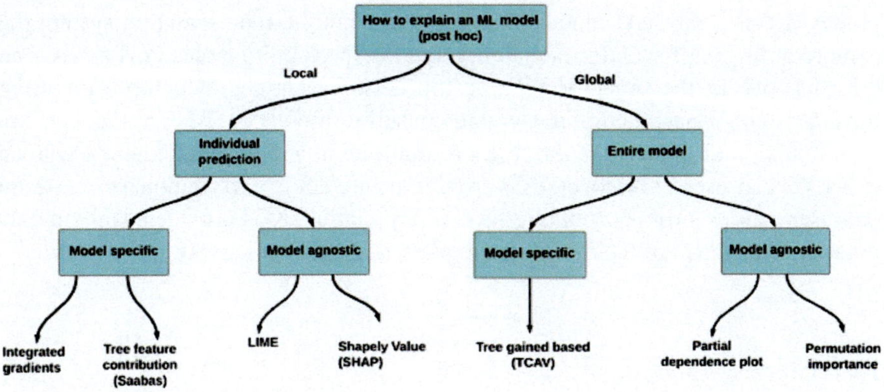

Fig. 6.2 Taxonomy of explainable AI

at its prediction. For this purpose, a simplified interpretable model is used to approximate smaller relevant aspects in a black-box system.

2. Global method: global approaches use the entire information about the model, the linked data, and the training to focus on the inside of the model. This gives us a bird's eye view of the model and how the various components of the data interact and affect the end result. In essence, it is an attempt to characterize the essence of the model.
3. Model-specific: model-specific interpretation methods are developed based on the parameters of specific models. Grad-CAM, for example, allows feature visualizations in the case of CNNs, but not in the case of LSTMs. These approaches often use feature maps formed by graph convolution, for example. Tree interpreters, for example, are determined by the type and functionality of the model.
4. Model agnostic: Model agnostic methods are explanatory methods that can describe any model and are not locked into a particular type of explanatory method. They are mostly used in post hoc analyzes and are therefore not bound to a specific model architecture. Internal model weights and structural factors are not directly accessible via the black-box model.

6.4 Explainable AI for Knowledge Management

Explainable AI (XAI) is a new research topic that aims to make AI systems more transparent and comprehensive [6]. Moreover, XAI leads to a divide between data scientists and knowledge scientists. Data scientists strive to collect and process uncontrolled and disconnected data without any context between the data. In contrast, knowledge scientists focus on semantic AI, which is an integral part of machine learning. Knowledge scientists neither extract normalized data that can be easily transformed into structures nor provide linked information by building knowledge graphs (KGs) as an XAI model. KGs are the symbolic representation system that is more promising for understandability and interpretability issues [7–10]. KGs are the backbone of the semantic relationship between entities, which are important for consistency analysis, inference, and causality inference. KGs are the semantic representation of knowledge that is more transparent and understandable to human experts or end-users. Moreover, KGs provide an interactive explanation for answering end-user queries through symbolic-level interpretation. KGs provide insight into the internal mechanism of the model and explain it more interactively.

References

1. Kute, D. V., Pradhan, B., Shukla, N., & Alamri, A. (2021). Deep learning and explainable artificial intelligence techniques applied for detecting money laundering—A critical review. *IEEE Access, 9*, 82300– 82317. https://doi.org/10.1109/ACCESS.2021.3086230
2. Lipton, Z. C., *The mythos of model interpretability*.
3. Heimerl, A., Weitz, K., Baur, T., & Andre, E. (2020). Unraveling ML models of emotion with NOVA: Multi-level explainable AI for non-experts. *IEEE Transactions on Affective Computing*, 1–1. https://doi.org/10.1109/TAFFC.2020.3043603
4. Wells, L., & Bednarz, T. (2021). Explainable AI and reinforcement learning—A systematic review of current approaches and trends. *Frontiers in Artificial Intelligence, 4*, 550030. https://doi.org/10.3389/frai.2021.550030
5. Meyes, R., de Puiseau, C. W., Posada-Moreno, A., & Meisen, T.: *Under the hood of neural networks: Characterizing learned representations by functional neuron populations and network ablations*. arXiv:2004.01254
6. Belle, V., & Papantonis, I. (2021). Principles and practice of explainable machine learning. *Frontiers in Big Data, 4*, 688969. https://doi.org/10.3389/fdata.2021.688969, https://www.frontiersin.org/articles/10.3389/fdata.2021.688969/full
7. Futia, G., & Vetrò, A. (2020). On the integration of knowledge graphs into deep learning models for a more comprehensible AI—Three challenges for future research. *Information, 11*(2), 122. https://doi.org/10.3390/info11020122, https://www.mdpi.com/2078-2489/11/2/122
8. Guo, Q., Zhuang, F., Qin, C., Zhu, H., Xie, X., Xiong, H., & He, Q. (2020). A survey on knowledge graph-based recommender systems. *IEEE Transactions on Knowledge and Data Engineering*, 1–1. https://doi.org/10.1109/TKDE.2020.3028705, https://ieeexplore.ieee.org/document/9216015/
9. Weis, J. W., & Jacobson, J. M. (2021). Learning on knowledge graph dynamics provides an early warning of impactful research. *Nature Biotechnology*. https://doi.org/10.1038/s41587-021-00907-6, http://www.nature.com/articles/s41587-021-00907-6
10. Kamal, M. S., Northcote, A., Chowdhury, L., Dey, N., Crespo, R. G., & Herrera-Viedma, E. (2021). Alzheimer's patient analysis using image and gene expression data and explainable-AI to present associated genes. *IEEE Transactions on Instrumentation and Measurement, 70*, 1–7.

Chapter 7
A Bibliometric Analysis of Artificial Intelligence in Knowledge Management

7.1 Introduction

AI enables machines to acquire, process, and use knowledge to perform tasks and unlock knowledge that can be passed on to humans to improve decision making [1–3]. AI and Knowledge Management (KM) are two sides of the same coin. KM enables insights into knowledge, while AI provides the ability to extend, use, and create knowledge in ways we have not yet envisioned. Artificial intelligence (AI) and its rapid technological advancements will significantly affect the future of work and how organizations manage their knowledge management (KM) processes. In this chapter, we have tried to carry out bibliometric analysis and co-occurrence analysis of Artificial Intelligence in KM to understand future trends of research in this area.

7.1.1 Bibliometric Analysis of AI in Knowledge Management

Data were retrieved from Scopus [4] database between January 1, 2000, and Oct. 30, 2021. The search term in both the cases was the keyword "Knowledge Management" and "Artificial Intelligence" in the title, abstract, and keywords (search option (TITLE-ABS-KEY ("Knowledge Management System") AND TITLE-ABS-KEY ("Artificial Intelligence"))). The data were analyzed using VOS viewer and CSV dataset format. The consistency and reliability of the data were checked (e.g., lack of consistency in country names and titles that sometimes abbreviations, acronyms, etc.) before bibliometric analysis [5].

Based on a search with the keyword "Knowledge Management" and "Artificial Intelligence" between January 1, 2000, and Oct. 30, 2021, the results showed 192 documents. It is interesting to note that in the area of "Knowledge Management" and "Artificial Intelligence" research, a sharp increase has been yet observed (Fig. 7.1).

62.5% results of our studied documents are conference papers and 27.1% results are articles; 3.1% is book chapters, and only 1% is books (Fig. 7.2). It is interesting

S. Majumder and N. Dey, *AI-empowered Knowledge Management*,
Studies in Big Data 107, https://doi.org/10.1007/978-981-19-0316-8_7

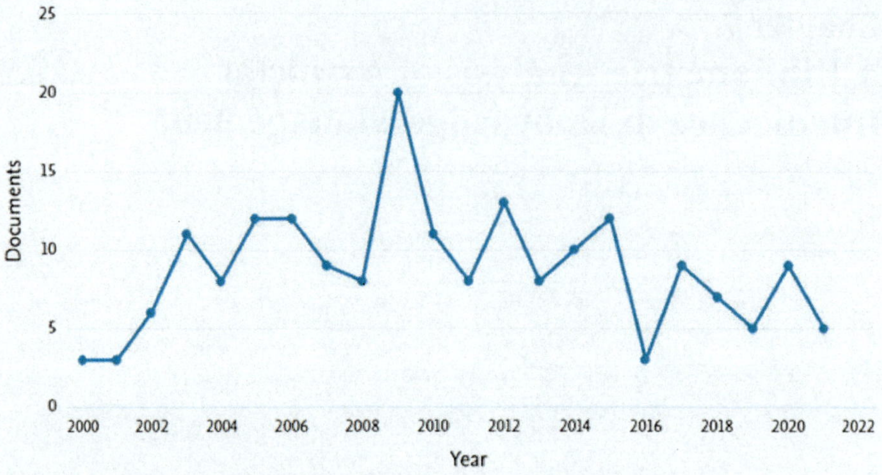

Fig. 7.1 Documents by year in Scopus

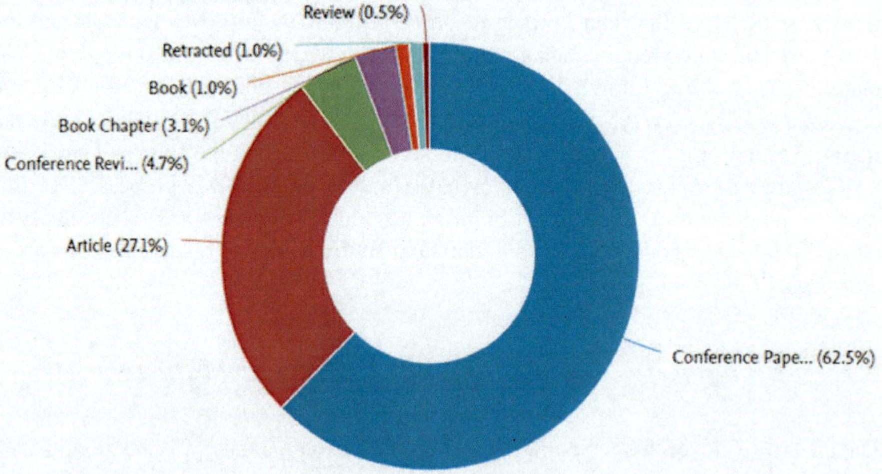

Fig. 7.2 Documents by type in Scopus database

to notice that majority of the work has been administered within the area of computer science (45%) and engineering (18%) (Fig. 7.3).

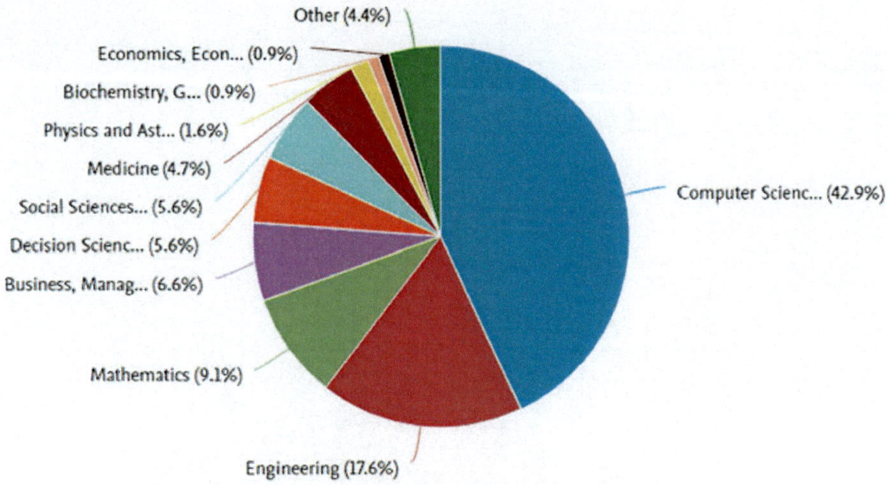

Fig. 7.3 Documents by subject area in Scopus database

7.1.2 Co-occurrence Analysis of Artificial Intelligence in Knowledge Management

In this chapter, we have also tried to find out the trend of Artificial Intelligence in Knowledge Management-related research. Total 192 documents were published till date (30.10.2021) where Artificial Intelligence and Knowledge Management has been mentioned in the title, abstract, keywords, and listed in the Scopus database. We have used VOS viewer software to analyze the database (word co-occurrence map out of text data).

From 192 documents, VOS viewer software extracted 45 keywords in titles (items) for network visualization. Those 45 items are grouped into three clusters (total link 891and link strength 2343) where each of the clusters contains a different number of items (Fig. 7.4). A link means a co-occurrence connection between two items. According to the VOS viewer manual, each link has strength, represented by a positive numerical value. The higher this value, the stronger the link. The total link strength indicates the number of publications in which two items occur together.

In cluster 1, there are total of 22 items (area, article, challenge, context, decision support system, dss, evolution, field, impact, implementation, kms, knowledge management systems, organization, person, perspective, role, state, strategy, type, variety, view) (Fig. 7.5). Among all these items, total link strength is the highest for "organization" (205). Links and occurrences are 43 and 42, respectively.

In cluster 2, 14 items are agent, case, content, data, document, goal, internet, ontology, proceeding, reasoning, technique, topic, use, and web, respectively (Fig. 7.6). In these cluster items, total link strength is the highest for "use" (86). Second and third highest are "data" (166) and "technique" (150), respectively. We have chosen "use" for visual representation (Fig. 5.6).

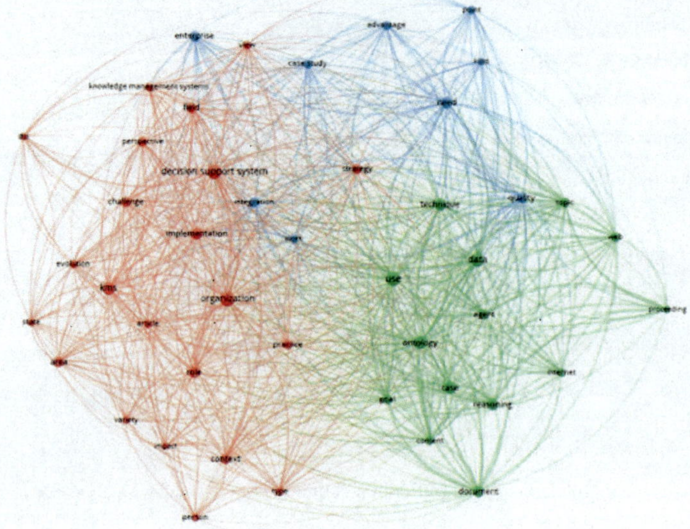

Fig. 7.4 Visualization of titles co-occurrences network

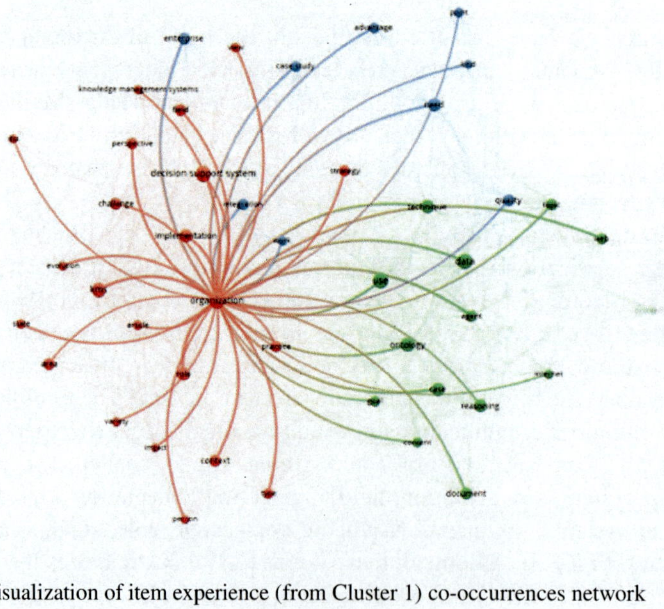

Fig. 7.5 Visualization of item experience (from Cluster 1) co-occurrences network

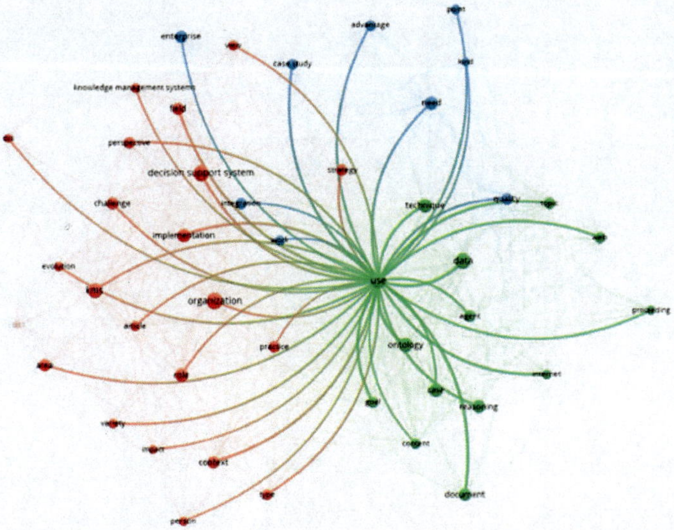

Fig. 7.6 Visualization of item experience (from Cluster 2) co-occurrences network

In cluster 3, there are nine items namely: advantage, case study, enterprise, integration, kind, need, point, quality, and work (Fig. 7.7). Among all these items, total link

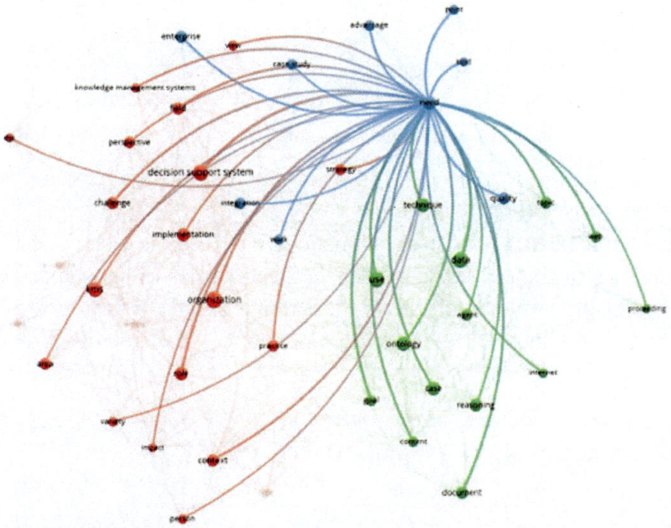

Fig. 7.7 Visualization of item experience (from Cluster 3) co-occurrences network

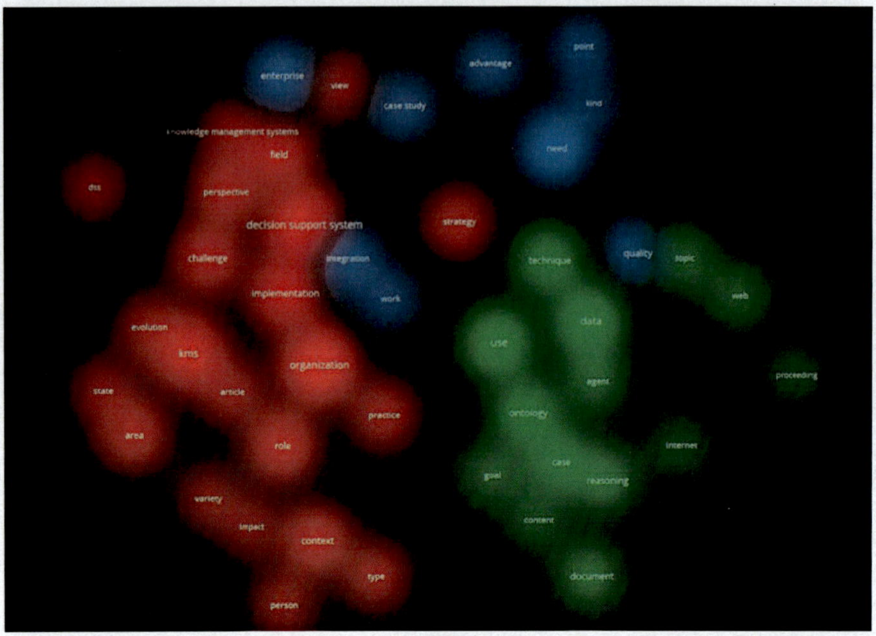

Fig. 7.8 Density visualization of title co-occurrences network

Table 7.1 The maximum occurrences of substance (Item) of each cluster

Items	Links	Total link strength	Occurrences
Organization	43	205	42
Use	43	186	35
Need	40	144	27

strength is the highest for "need" (144). Second and third highest are "integration" (99) and "enterprise" (93), respectively.

Figure 7.8 indicates the density visualization based on the total hyperlink strength. The dimensions of the circles represent the occurrences of substances (items). We have looked at the circles with the highest range based on the sizes of the circles from each cluster. Clusters 1, 2, and 3 are mentioned, respectively (Table 7.1).

7.2 Trend Analysis in Decent Work Research

From this study, we have come to the conclusion that the overall link (co-incidence connection among objects) strength shows the quantity of publications wherein items occur together (Table 7.2).

Table 7.2 Organization in AI-empowered knowledge management based study

Cluster number	Items	Links	Total link strength	Occurrences
1 (22items)	area, article, challenge, context, decision support system, dss, evolution, field, impact, implementation, kms, knowledge management systems, organization, person, perspective, role, state, strategy, type, variety, view	43	205	42

Table 7.1 clearly shows that "organization" having higher link strength (205) than the rest of the items but interestingly once cluster two is analyzed; it is observed that "decision support system" contributed 175 total link strength. Hence, we can conclude that co-occurrence analysis of "Artificial Intelligence in Knowledge Management" literature published from the year 2000 to end of Oct. 2021 clearly shows that organization plays a significant role in decent work-related studies.

7.3 Conclusions

In this chapter, our main objective was to analyze the bibliometric and co-occurrence of Artificial Intelligence in Knowledge Management to understand trends of research in this area and also find out the scope of the same. Two important findings from bibliometric and co-occurrence analysis of AI-empowered knowledge management are (I) there are very limited books published which covers AI-empowered knowledge management and (II) Such type of research plays a significant role in organizations and lots of opportunities will be created in future to contribute in this domain.

References

1. Fong, S. J., Dey, N., & Chaki, J. (2021). *Artificial intelligence for coronavirus outbreak*. Springer.
2. Dey, N., & Ashour, A. S. (2017). Ambient intelligence in healthcare: A state-of-the-art. *Global Journal of Computer Science and Technology.*
3. Bhatt, C., Dey, N., & Ashour, A. S. (Eds.). (2017). *Internet of things and big data technologies for next generation healthcare.*
4. https://www.scopus.com/ (Last access date: June 30, 2021).
5. Al-Zaman, M. (2021). A bibliometric and co-occurrence analysis of COVID-19–related literature published between December 2019 and June 2020. *Science Editing, 8*(1), 57–63.

Chapter 8
Conclusions

The main focus of this book is knowledge management in the organization through artificial intelligence. Knowledge management helps the workforce and the organization both for collaborating, sharing, creating, and using knowledge in its best way. A proper understanding of knowledge management system improves performance of the business, increases innovation and creativity, and develops knowledge base of the workforce and the organization. The knowledge that is applied in decision-making process of an organization is accurate, dynamic, and personal in nature. Artificial intelligence has played a great role in this scenario. AI with machine learning helps to allow machines to acquire processes and use knowledge for better performance of manpower and firms. By unlocking knowledge it enhances the quality of decisions under decision-making process. AI maintains an important role to deliver knowledge in a digitized organization. Artificial Intelligence-based knowledge management enables people to acquire knowledge regarding activities. AI works toward scaling the volume and effectiveness of knowledge distribution. Artificial intelligence is extensively used in the process of decision making without taking support from humans or with very less intervention. The advantage of this system is that in this case, knowledge is unbiased, and decisions that are made with knowledge, and are ethical in true sense. AI enables faster, accurate, and efficient decision-making techniques in the process of delivering knowledge. AI-based knowledge management has an unimaginable capability to expand, use and generate knowledge. AI-empowered knowledge management practice in the organization can detect patterns in enormous volumes of data and information. In recent years, progress in AI schemes helped the KM to reach its next level. Recent advancements, such as information optimization, IoT, and cloud computation enhanced the existing KM systems in the industries. On implementing AI-based KM, handling of the information sharing process is automated, which reduces the data handling burden that existed in the earlier KM scheme. AI with machine learning algorithms is very much used for delivering and disseminating knowledge throughout the workplace. Keeping in mind the importance of artificial intelligence in paradigm of knowledge management, businesses should focus on responsible use and responsible design of AI.

S. Majumder and N. Dey, *AI-empowered Knowledge Management*,
Studies in Big Data 107, https://doi.org/10.1007/978-981-19-0316-8_8

The main objective of knowledge management in a business is to understand properly about the creation of knowledge, transfer of knowledge, use of knowledge, flows of knowledge, and finally governs of knowledge. The implementation of knowledge management and its efforts helps to achieve the mission of an organization. It improves activities of its end users. The use of tacit and explicit knowledge and their conversion also serves the vision statement of a firm. This enables better understanding of social network analysis within these firms. It is not always important to understand the effectiveness of knowledge management in a business house, the nature of organization's knowledge itself should be important at this juncture. Knowledge is dynamic and continuously it is updated under the process of knowledge management life cycle in a business organization. This process includes the experts who are providing insights about knowledge. The dynamic components of knowledge make the positive image and brand value of the organization, and it evolves over time. Knowledge Management System enables preserving productivity and saving time. The adaptability of team members to the knowledge management system is wider and encouraging. In this way, continuous upgradation and use of KMS is really incredible. The personalized components of knowledge answer the questions of users when they are in need. Personalized knowledge has been facilitated on the basis of flow of knowledge throughout the organization. Knowledge in the organization is identified as the source of truth. In this book, we have focused on different tools of knowledge management and how it works effectively in this business environment. Besides, we have discussed the devices currently implemented in knowledge management systems and the impact of artificial intelligence in the process of knowledge management. Nowadays, knowledge is so much powerful in achieving organization's long-term and short-term goals. The applications of knowledge management in some major working sectors like health care, construction, education, etc. are also discussed in this book. A premier and unique way of knowledge management through explainable AI is also portrayed in our book. Explainable Artificial Intelligence (XAI) is a transparent window that displays internal processing in machine learning models. Explaining ability, in this case, means understanding how a model works, but not in a complex manner that is difficult for an average person to understand. We have also reported a chapter on bibliometric and co-occurrences analysis which helps to investigate the scope of research in the area of artificial intelligence in knowledge management. Two important findings from bibliometric and co-occurrence analysis of AI-empowered knowledge management are: (I) there are very limited books published that covers AI-empowered knowledge management and (II) such type of research plays a significant role in organizations and lots of opportunities will be created in future to contribute to this domain.